Chimera's Children

This book was commissioned by the Scottish Council on Human Bioethics (SCHB). The Council was formed in 1997 as an independent, non-partisan, non-religious body made up of physicians, lawyers, ethicists and other professionals from disciplines associated with medical ethics. The principles to which the SCHB subscribes are set out in the United Nations Universal Declaration of Human Rights which was adopted and proclaimed by the UN General Assembly by resolution 217A (III) on 10 December 1948.

Scottish Council on Human Bioethics, 15 Morningside Road, Edinburgh EH10 4DP, SCOTLAND Tel: 0131 447 6394; E-mail: mail@schb.org.uk; www.schb.org.uk

Chimera's Children

Ethical, philosophical and religious perspectives on human-nonhuman experimentation

**EDITED BY
CALUM MACKELLAR AND
DAVID ALBERT JONES**

continuum

Continuum International Publishing Group

The Tower Building	80 Maiden Lane
11 York Road	Suite 704
London	New York
SE1 7NX	NY 10038

www.continuumbooks.com

First published 2012

British Library Cataloguing-in-Publication Data
A catalogue record for this book is available from the British Library.

ISBN: 978-1-4411-6984-6 (hardback)
978-1-4411-9886-0 (paperback)

Library of Congress Cataloging-in-Publication Data
A catalog record for this book is available from the Library of Congress.

Typeset by Fakenham Prepress Solutions, Fakenham, Norfolk NR21 8NN
Printed and bound in India

Contents

Authorship

The book has been the work of many hands. Most of the initial drafting has been done by the principal editor, Dr Calum MacKellar with major contributions from Dr David Albert Jones. In addition, Prof. Damien Keown drafted the section on Buddhist perspectives with Dr Sibtain Panjwani and Mr Imranali Panjwani preparing the section on Islamic perspectives.

Acknowledgements

The book is indebted to the following (though it should not necessarily be assumed that they would endorse all of its contents): Dr Elizabeth Allan (Biologist: Member of the Scottish Council on Human Bioethics); Mrs Ann Bruce (ESRC Centre for Social and Economic Research on Innovation in Genomics: University of Edinburgh); Dr Neville Cobbe (Cell Biologist: University of Edinburgh); Dr Kenneth Collins (General Practitioner: Member of the Scottish Council of Jewish Communities, Glasgow); Mr Philip Hannan (Civil Servant in Scotland); Dr Matthew Harvey (Research Fellow with ESRC Genomics Policy and Research Forum: University of Edinburgh); Dr Susan Holloway (Clinical Scientist: Clinical Human Genetics, Molecular Medicine Centre, Western General Hospital, Edinburgh); Sheikh Arif Abdul Hussein (Director of Al-Mahdi Institute, Birmingham); Dr Rhona Knight (General Practitioner and Biomedical Ethicist: Member of the Nuffield Council on Bioethics); Dr David King, (Biologist and Biomedical Ethicist, Member of Human Genetics Alert); Dr Gregory K. Pike (Director, Southern Cross Bioethics Institute, Australia); Mr Adam Ransom (Theology Department: University of Chichester); Ms Hanne Stinson (Chief Executive, British Humanist Association, London); and Dr Agneta Sutton (Biomedical Ethicist: Associate Lecturer at the University of Chichester).

Preface

This book takes its title from a creature of ancient mythology, 'The raging Chimera, she was of divine stock, not of men, in the fore part a lion, in the hinder a serpent, and in the midst a goat, breathing forth in terrible wise the might of blazing fire' (Homer the *Iliad* 6/180). The quotation is from Homer's *Iliad* but evidence exists from earlier Greek pottery that the Chimera and other admixed creatures haunted the human imagination even before written history. The Chimera is of godly origin being offspring of Echidna, the mother of monsters (see Hesiod *Theogony* 319–25). She is powerful and unnerving precisely because she is made of different parts of different animals – lion, snake and goat.[1] None of the Chimera's parts is human but there are other legendary creatures where the human is part of the mixture like the minotaur (part man and part bull). Such myths of monsters performed various functions, but most notably they explored the idea of a beast within and untamed emotion: the violence of the werewolf or the minotaur; the sensuality of the faun and the mermaid. A variant of this theme is revived in H. G. Wells' *The Island of Doctor Moreau*. The Chimera and these other monsters were also symbolic of some frightening disorder threatening both the universe and humanity with chaos (Karpowicz et al. 2005).

Evoking the Chimera of classical myth is relevant because that is where the modern scientific term originates, but it functions partly by way of contrast with the issues presented in this book. Interspecies admixed organisms, including part human admixed organisms, are not only a feature of ancient mythology. They represent a real scientific possibility and in some cases, an actuality that is already present.

In a series of dramatic experiments, which took place at the end of the twentieth century, small brain sections from developing quails were taken and transplanted into the developing brains of chickens which, eventually, began to exhibit vocal trills and head bobs unique to quails. This demonstrated that complex behaviours could be transferred from one species to another and emphasised the potential power of embryonic and foetal interspecies combinations (Balaban 1997, Weiss 2004).

---

[1] The chimera was killed by the hero Bellerophon on the winged horse Pegasus (Pearsall and Trumble 1996).

Because of these findings and others like them, the President's Council on Bioethics of the USA indicated, in a report published in 2004, that in the context of actually mixing human and nonhuman gametes or blastomeres (a single cell in a very early embryo), the ethical concerns raised by violating the human-nonhuman species barrier were especially acute. In particular, the President's Council recommended that a bright line should be drawn at the creation of human-nonhuman embryos produced by the fertilisation of human eggs by the sperm of nonhuman animals (for example, chimpanzee), or the reverse. The Council recommended that the US Congress should, therefore, draft legislation to address these biological possibilities and make it illegal to cross this line. This is a position which has, so far, not been altered in the USA, since no moves have been proposed to grant federal funding to create human-nonhuman embryonic and foetal combinations for research even though public policy on human embryonic stem cells has been liberalized.

In stark contrast, however, the UK House of Commons Science and Technology Committee (HCSTC) prepared a report in 2005 which took a very different position from the US President's Council. It recommended that the fertilisation of nonhuman eggs with human sperm should continue to be legal and state-funded in the UK for research purposes (HCSTC 2005: 30–2). In addition, it indicated that the time limit, before which they must be destroyed, should be extended. The same House of Commons Committee re-emphasized this permissive position with its 2007 report by indicating that:

We believe that there is a need to allow research using some forms of human-animal chimera or hybrid embryos, including but not exclusively cytoplasmic hybrid embryos, *to proceed immediately* ... We believe that, in general, the creation of *all types of human-animal chimera or hybrid embryos* should be allowed for research purposes ... [emphasis added]. (HCSTC 2007: 62–2)

There is thus no international consensus on whether to permit or to promote research on human-nonhuman embryonic or foetal combinations. Furthermore, the evidence suggests that in most developed nations public opinion is divided on the issue, with the balance in favour of setting strict limits to this kind of research (BBVA 2008: 28). The issues are complex and in most countries there has been little public debate (Hug 2009). When the possibility of human-nonhuman combinations has been reported in the media there has been a tendency, on the one hand, to hype the prospective benefits and on the other hand to play on irrational fears of possible dangers. In reality, however, there are many subtly different kinds of human-nonhuman combinations and many cultural, ethical and world view perspectives which need to be taken into account.

It may be that reflecting on ancient myths in this area can help inform ethical and philosophical reflection (for ethical and philosophical truths are often conveyed by myths). Nevertheless, whether or not this is true, it is necessary first of all to set aside the fabulous imagery and be clear what is meant or implied by 'chimera' in a modern research setting. In this context, because the chimera represented a combination of different parts originating from different species, the term has actually been appropriated by biologists to broadly describe any biological entity resulting from a combination of materials from two or more different organism (Greely 2003).

The chimeras of modern science are thus very different from the monsters of myth. Monsters were typically regarded as the result of divine intervention, often malign. In contrast chimeras, hybrids and transgenic organisms in bioscience are generally the result of human intervention, using and presupposing the biological powers of nature. Such activities are not arbitrary but are deliberate actions in the pursuit of various goals.

The present book is offered as a clarification of some of the questions that are pressing and perplexing in equal measure. The book is divided into three parts. The first part looks back from the myths of ancient Greece through recent science fiction to contemporary science and legal regulation. It demonstrates that while the science may be new, the idea of human-nonhuman combinations is ancient. The second part looks towards the future development of different kinds of human-nonhuman combinations. It seeks to move away from science fiction and consider in more detail those combinations that are realistic in the future and those that have already been created. The third part looks at contemporary cultural, worldview and ethical perspectives. It aims to give depth to the current discussion by broadening the context for debate. The topics covered by different chapters (for example, Chapter 2 on legal aspects or Chapter 9 on worldview perspectives) can generally be read independently of one another and the book can be used in this way, as a resource. However the aim of the book is more than this. It seeks to understand the issue synoptically, applying to these novel scientific possibilities principles distilled from a variety of perspectives. There are no easy answers here, but this discussion is offered as a starting point for deeper reflection.

Calum MacKellar
David Albert Jones

January 2012

PART ONE

Background, definitions and current legislation

Introduction

The idea of crossing the human species barrier has always fascinated humanity. In ancient Egypt, Greece and Rome, for example, mythological creatures such as sphinxes (often with human-lion combinations), centaurs (human-horse combinations), fauns (human-goat combinations), harpies (bird-human combination) and the minotaur (a human-bull combination, the mythical result of a sexual union between the Queen of Crete, Pasiphae, and a bull) were portrayed both in legend and in art as being special and endowed with extraordinary powers (McLaren 1976: 1).

Interestingly, although such beings were not considered as being part of the human race, neither were they seen as being inhuman or completely bestial, their distinct and solitary status in mythology sometimes resulting in them being rejected as different and portrayed as lonely monsters. Indeed, Western historical and cultural beliefs often considered the mixing of two separate categories of animals as representing an evil (Bazopoulou-Kyrkanidou 2001). For example, the myth relating to the minotaur portrays the monster as dangerous and malign necessitating its eventual destruction by Theseus. Similarly, pictorial representations of the devil and demons in medieval times include both human but also animal parts such as horns and tails.

One enduring story which was taken seriously for many centuries was the existence of a cynocephalic (dog-headed) race. They were described as having the head of a dog with a human body. They ate human flesh and barked. According to Herodotus (484–425 BC), these creatures lived in the forests of Libya (*History* 4.191). On the other hand, Ctesias, another Greek travel writer and historian of the same period, wrote that cynocephali were to be found in India (as cited in a fragment preserved by Photius (c.815–897)). The same story was still around eight centuries later and is mentioned by Jerome (347–420) (Gordon White 1991). Thomas of Cantimpré cites Jerome on the existence of Cynocephali, in his *Liber de Monstruosis Hominibus Orientis*, xiv. Augustine (354–430) was not convinced that this race actually existed, but was concerned to understand the implications should such a species be discovered.

> I must therefore finish the discussion of this question with my tentative and cautious answer. The accounts of some of these races may be

completely worthless; but if such people exist, then either they are not human; or, if human, they are descended from Adam. (*City of God* 16.8)

After Augustine few writers showed his 'tentative' caution. The story of a race of dog-headed people seems to have been widely believed in the early middle ages. Medieval travellers John de Plano Carpini (c.1182–1252) and Marco Polo (1254–c.1324) both mention cynocephali in their writings with the former mentioning a race of dog-heads which lived north of what may be recognised as Lake Baikal.

In a further twist, the myth of the dog-headed race was fused with the legend of St Christopher who lived in the third century. It is said that he was a cynocephalic convert who, on receiving baptism, was rewarded with a human appearance. This story occurs in sources from the fourteenth century and probably existed much earlier. There are even some icons in Eastern Orthodox Churches in which St. Christopher is portrayed with a dog head.

It is interesting that the myth of the cynocephali has consistent features from Herodotus to the medieval legends of St Christopher. It frequently associates dog-headedness with savagery and cannibalism. The dog-headed person is a human being reduced to the savagery of the wild animal. They are the 'wild men of the woods', a kind of mythical 'missing link' between beast and man. The medieval idea that dog-headed races existed in remote corners of the world is directly parallel with the modern legends of the yeti or Bigfoot: an elusive semi-human creature who lives in wild places and who provides an image of humanity before civilisation.

These stories, like similar legends of mermaids and werewolves, have been common in many cultures. They reflect an enduring fascination with creatures who mix human and nonhuman characteristics while raising questions about the importance of the human body and human identity (Haddow et al. 2009).

In at least some cases, these mythical semi-human figures seem to be based on human biological dysfunction wrongly assumed to be due to human-nonhuman hybridization. For example, a man might mistakenly be considered to be a werewolf because hair covered his whole body due to a rare genetic condition (Leroi 2005: 268 ff.).

The description of human medical conditions as belonging to a nonhuman form of animal life can still be found in a book written as recently as the 1950s. In *The Sanctity of Life and the Criminal Law* Glanville Williams applies the term 'monsters' to children born with severe disability.

On rare occasions such a monster will live. It may belong to the fish stage of development, with vestigial gills, webbed arms and feet, and sightless eyes. The thing is presented to its mother, who struggles to nurture it for a few months, after which she sends it to a home. (Williams 1958: 33 on webbed feet see Leroi 2005: 117)

In this regard, a 'monster' would be a being that disturbs and challenges the settled boundaries of nature and which make sense to society (see Jasanoff 2005).

Williams concluded that it was probably lawful to kill this kind of human 'monster'. He claimed that this conclusion was supported by the medieval jurist Bracton, though he admitted that Bracton's opinion was based on the mistaken belief that such progeny were the result of intercourse with animals. Williams argued that 'the same rule might be approved for a better reason', by which he meant the eugenic reason 'of preventing the birth of children who are congenitally deaf, blind, paralysed, deformed, feeble minded, mentally diseased, or subject to other serious hereditary afflictions' (Williams 1958: 82; Keown and Jones 2008). It is extraordinary that these words could be written by a respected English academic only ten years after the Nuremburg trials following the Second World War. Here the myth of the existence of not-quite-human beings seems to endure as a denial of the fundamental moral equality of people with disability.

In addition to the misuse of hybrid imagery by modern eugenicists, another modern mode of discourse on human-nonhuman interspecies entities is the medium of science-fiction. These works have frequently portrayed the difficulties of defining the human versus the nonhuman elements of interspecies beings.

For example, H. G. Wells' (1866–1946) book *The Island of Doctor Moreau*, published in 1896, tells the story of Edward Prendick who is shipwrecked and ends up on a desert island: a place where a certain Dr Moreau is undertaking vivisection experiments to turn nonhuman animals into humans. At first revolted, Prendick later comes to empathize with these created beings, who have the ability to speak, but he is eventually confronted with questions of humankind's own animalistic instincts. The novel allows the exploration of the animal side of human nature and of the implications of Darwin's claims about the origin of man. Moreover, the story explores the idea and implications of the beast-man in addition to examining the hubris of Dr Moreau, the scientist who is driven to pursue his research wherever it leads at whatever cost and who has 'never troubled about the ethics of the matter' (Wells 2005: 75).

A novel which studies the very difficult problems of identifying the moral status of human-nonhuman interspecies beings was written by the French author Vercors (whose real name was Jean Bruller (1902–1991)). Published in French in 1952 as *Les Animaux dénaturés*, it appeared in English in 1976 under the title *Borderline*. The book tells the story of a team of anthropologists looking for the 'missing link' between humans and nonhuman primates and who actually discover such a species in New Guinea which they name Tropi. The discovery, however, creates a storm which threatens to disrupt the set thinking of twentieth-century society because a creature on the borderline between humans and nonhuman primates seems to undermine the privileged moral status of human beings.

When a very large textile company seeks to use the tropis as slaves in its factories, the hero of the novel, an English journalist named Douglas Templemore, decides to artificially inseminate a female tropi with his sperm before bringing her back to London and killing the resulting child. This he does in order to force the English legal system (and the UK Parliament) to decide whether or not he has committed a murder. In so doing, the hero would determine, once and for all, the moral status and rights of the tropi. After many intractable arguments, however, the English authorities eventually decide not to attempt to answer the question whether or not a tropi is human until the more basic question of the nature of humankind is defined. But, to the consternation of all, it is discovered that a definition of humankind has never been determined or established in English law. The question 'What is humankind' may have been one which Vercors asked himself when confronted with the racial discrimination of the Nazis towards different kinds of human beings during the Second Wold War in which he was a resistance fighter in the Vercors region of France.

The theme of intelligent nonhumans is also explored in another French science fiction novel: Pierre Boulle's *La Planète des singes* (1963). This was the basis of the highly successful film franchise, *The Planet of the Apes* (1968, with sequels in 1970, 1971, 1972, 1973, 2001 and 2011). In the book and in the first series of films, the transformation of the apes is the result of training. However, in the latest film, it is a result of a genetically engineered retro-virus, with the implication it may have introduced human genes into the apes. A similar move is made in *Next,* the 2006 techno-thriller by the American author Michael Crichton. He imagines a humanzee being created through the introduction of human genes into the genetic constitution of a chimpanzee. In this story the humanzee has distinctly human characteristics and Crichton uses the plot to reflect his own perspective and concerns relating to new genetic experimentation.

Such stories are important and not entirely unbelievable because the process of interspecies mixing can and does occur when species are genetically very close. The most common and well-known example of a nonhuman animal-animal combination is the mule. This is a true hybrid. In other words, fertilization of a horse egg with donkey sperm produces an animal whose every cell contains chromosomes from both parental species. Conversely, a hinny is obtained when a male horse mates with a female donkey. Donkeys have 62 chromosomes, while horses have 64. Their offspring have 63 chromosomes which cannot evenly divide. Thus mules and hinnies are almost always sterile. There are also other examples of success in crossing animals of different species: a lion with a tiger; a goat with a sheep; a wolf with a dog. Occasionally this also happens without human intervention. In May 2006, DNA tests confirmed that a bear shot in the Canadian arctic was a hybrid of a polar and a grizzly bear. Such a cross (the 'grolar bear') had been known in captivity but the 2006 case was the first confirmed example in the wild. Combinations between biological species are thus relatively rare in nature, and most such entities are less 'fit' than their progenitors.

With respect to human-nonhuman combinations, no evidence of any entities being born has ever been recorded but new developments in assisted reproductive technology, genetics and biochemistry may well make this technically feasible. Human-nonhuman combinations are no longer confined to the domain of mythology but have become a possible object of scientific research. For example, in the mid-1920s the Soviet Union's top breeding scientists Prof Ilya Ivanov tried to impregnate female chimpanzees with human sperm in order to create a human-chimpanzee hybrid (a humanzee). These experiments were unsuccessful but were believed to be feasible by some of the leading scientists of the day (Rossiianov 2002). It was only because of grave ethical concerns expressed by the general public and other commentators at the time that this line of research was eventually abandoned. While such experiments were conducted by Soviet scientists they were also considered but rejected by that other totalitarian movement, German National Socialism. Adolf Hitler expressed indignation about the possible hybridization between humans and nonhumans, believing that the combinations always lead to degeneration (Rossiianov 2002: 310). Despite a lack of sympathy with many aspects of Christianity, Hitler wrote in *Mein Kampf* that: 'The state is called upon to produce creatures made in the likeness of the Lord and not create monsters that are a mixture of man and ape' (Hitler 1939 Vol. 2 Section 2).

More recently, new procedures have been developed by scientists which combine human and animal biological elements to such an extent that it questions the very concept of being entirely human (Robert 2006: 838–45). In the light of this research, concerns about human-nonhuman combinations were raised in 2001 by the UK Animal Procedures Committee. This indicated that, in addition to questions about the fate of such human-nonhuman combinations, a deep repugnance may exist at the thought of their very creation. Indeed, the committee indicated that the main opposition to human-nonhuman combinations would probably arise from 'those who wish to maintain real boundaries between human and nonhuman animals, and who retain a conviction that "kinds" are separate creations, each – as it were – designed to embody a particular beautiful form'. Thus, confusion of 'kinds' may be something which raises concern in a large section of society even though no certainty exists as to the exact identity of these 'kinds' (Animal Procedures Committee 2001: 18–20).

Generally, the term 'kinds' is not so much used in biology as in folk-taxonomy. From a biological perspective, living organisms are classified not into kinds but into species (that is, into groups which do not normally interbreed). Nevertheless, in moral and political discussion it is common to speak of human beings as belonging to the same 'kind'. This is important, for example, for the concept of universal human rights. It is unlikely that the idea of a shared human kind or a common human nature would be undermined if it were discovered that some human beings technically belonged to a different biological species (that is, that members of this population were not able to breed successfully with other human populations and vice versa). This would be biologically interesting but would not, normally, undermine their claim to equal respect as human beings (Animal Procedures Committee 2001: 18–20).

Moreover, there is considerable debate about the reality of biological species or whether they are the product of humanly imposed classification schemes which are empirical and pragmatic. For instance, one can classify species into groups in which organisms reproduce sexually if this mode of propagation is deemed to be the most significant criterion. But this raises the question of whether bacteria, because of their ability to transfer genetic material between themselves through cell-to-cell contact, can or should be divided into species, and whether reproduction criteria provide the most compelling reasons to draw lines between organisms. It is arguable that this and other classification schemes are sometimes formulated based on subjective human interests (Karpowicz, Cohen and van der Kooy 2004: 331–5, De Sousa 1980). This would mean that species categories were never real, ontological entities or natural kinds (De Sousa 1984). Nevertheless, while the category of biological species may not be universal or immutable it does serve a useful function and is hard to jettison altogether. Boundaries may be blurred at the edges but they still help to identify different characteristic forms of life – characteristic physiology, characteristic behaviour, characteristic place in the ecology. In the case of the human species these shared specific charac-teristics have moral and political significance.

On the other side of the Atlantic, the President's Council on Bioethics of the USA considered the topic in 2004. It concluded that crossing the human-nonhuman boundary was, in some respects, quite complex and subtle but that the combination of human and nonhuman tissues and materials was not in itself objectionable.

This means that, in the context of therapy and preventive medicine, the President's Council accepted that the transplantation of nonhuman animal parts to replace defective human ones could be considered as ethical.

Moreover, the Council had no specific objection to the insertion of nonhuman-derived genes or cells into a human body – or even into human foetuses – where the aim would be to address a serious disease in the patient or the developing child.

Similarly, in the context of biomedical research, the US Council did not oppose the practice of inserting human stem cells into nonhuman animals. However, in the context of procreation – of actually mixing human and nonhuman gametes or embryonic cells at the very earliest stages of their development – the Council indicated that the ethical concerns raised by crossing this boundary were especially acute. Thus, the drawing of clear lines limiting permissible research should be specifically considered.

Consequently, the President's Council recommended that one bright line should be drawn at the creation of human-nonhuman embryos, produced by the fertilization of human eggs by nonhuman animal (for example, chimpanzee) sperm or the reverse. This is because the Council accepted that society should not be asked to judge the humanity or moral worth of such ambiguous hybrid entities (for example, a 'humanzee', the analogue of the mule). Furthermore, the Council made it clear that it did not want to leave open the possibility of a human being having other than human progenitors (for example, having a monkey as a parent). The Council, therefore, recommended that the US Congress should draft legislation to address these biological possibilities and make it illegal to cross this line.

However, a more permissive approach to the creation of human-nonhuman combinations, especially in the realm of the embryo, was considered in a report entitled 'Human Reproductive Technologies and the Law' prepared in 2005 by the UK House of Commons Science and Technology Committee. In this report it is somewhat surprisingly indicated that,

> while there is revulsion in some quarters that [human-nonhuman] creations appear to blur the distinction between animals and humans, it could be argued that they are less human than, and therefore pose fewer ethical problems for research than fully human embryos.

The report went on to recommend (HCSTC 2005: 30–2) that new legislation should:

a) Define the nature of these creations,

b) Make their creation legal for research purposes if they are destroyed in line with the current UK 14-day rule for human embryo cultures, and

c) Prohibit their implantation in a woman.

This position was re-emphasized by the same House of Commons committee in its 2007 report entitled 'Government Proposals for the Regulation of Hybrid and Chimera Embryos' in which the Members of Parliament stated that:

> We believe that there is a need to allow research using some forms of human-animal chimera or hybrid embryos, including but not exclusively cytoplasmic hybrid embryos, *to proceed immediately* ... We believe that, in general, the creation of *all types of human-animal chimera or hybrid embryos* should be allowed for research purposes... . (HCSTC 2007: 62–3, emphasis added)

As a result of these scientific developments in the creation of human-nonhuman combinations, the Scottish Council on Human Bioethics prepared an extensive report, in 2006, on the ethical issues associated with human-nonhuman interspecies embryos. This was one of the first reports in Europe on the subject and sought to emphasize the ethical complexity and serious risks relating to such experimentation. However, not all commentators were prepared to be so categorical in their assessment of human-nonhuman interspecies embryos. Bioethicists such as Jason Scott Robert and Françoise Baylis for example, were unwilling to draw a specific line concerning such interspecies combinations (Robert and Baylis, 2003, Robert, 2006). They asserted that they took 'no stance at all' on whether 'animal-human mixtures should be forbidden or embraced' asserting that 'the arguments against ... creating novel part-human beings ... are largely unsatisfactory'.

This was reflected by Hugh McLachlan, a professor of applied philosophy in Scotland, who indicated that although the idea was 'troublesome', he could not see any ethical reasons to oppose the creation of humanzees. He argued that 'If it turns out in the future there was fertilisation between a human animal and a non-human animal, it's an idea that is troublesome, but in terms of what particular ethical principle is breached it's not clear to me', adding 'I share their squeamishness and unease, but I'm not sure that unease can be expressed in terms of an ethical principle' (Haworth 2008).

Finally, some commentators, such as zoologist Prof Richard Dawkins have even admitted to a 'frisson of enjoyment' at the prospect of having to question the hitherto unquestioned issue of the creation of human-nonhuman interspecies entities. He acknowledged at the beginning of 2009 that the successful hybridization between a human and a chimpanzee would change everything, adding: 'Even if the hybrid were infertile like a mule, the shock waves that would be sent through society would be salutary ... It cannot be ruled out as impossible, but it would be surprising' (Rabderson and Dawkins 2009). In this regard, he also indicated that:

If there were a heaven in which all the animals who ever lived could frolic, we would find an interbreeding continuum between every species and every other. For example I could interbreed with a female who could interbreed with a male who could ... fill in a few gaps, probably not very many in this case ... who could interbreed with a chimpanzee. We could construct longer, but still unbroken chains of interbreeding individuals to connect a human with a warthog, a kangaroo, a catfish. This is not a matter of speculative conjecture; it necessarily follows from the fact of evolution. (Rabderson and Dawkins 2009)

Author, James Hughes, in his 2004 book, *Citizen Cyborg*, has also indicated that he would welcome the creation of a human/chimpanzee hybrid that could interbreed with both species, and thereby 'break the species' barrier. This, he wrote, would prove that humans are not special and undermine what he calls 'human racism'.

From the above comments, it is easy to see how the topic of human-nonhuman interspecies entities relates to some of the most fundamental questions facing humanity. Not only does it ask questions about the moral status of new biological beings but it also forces humanity to reconsider and re-evaluate itself in the context of these beings. In other words, the ethical considerations run very deep and are extremely complex. It was in order to seek some clarifications, where possible, relating to the scientific, philosophical, cultural and religious perspectives that the present volume was prepared.

1

Historical background

From a scientific perspective, attempting to create human-nonhuman embryonic combinations does not stretch back very far and only a few experiments or procedures have, until now, taken place in this field.

Early discussions

Scientific discussions relating to the creation of human-nonhuman inter-species entities began in 1908 with the publication of *Truth: Experimental Researches about the Descent of Man*, a short book by Marie Bernelot Moens, a college teacher and amateur zoologist from Maastricht, Holland. In this, he proposed to inseminate a chimpanzee female with human sperm in order to provide experimental conclusions in relation to evolutionary theory. In this regard, it was reported that Moens had actually discussed his project with the German professor of biology Ernst Haeckel, one of the most distinguished European experts on evolution at the time.

Quoting recent results from comparative blood studies, Haeckel apparently expressed the view that Moens experiments could result in the birth of a live hybrid entity, and that this would be invaluable for a better understanding of human evolution. Believing in the separate evolutionary origins of different human races, Haeckel also recommended that, for the experiment to be successful, Moens had to use the sperm of an African man. However, Moens attempts to organize an expedition to Africa in order to capture chimpanzees for his experiments resulted in a scandal which eventually cost him his teaching job (Rossiianov 2002: 291).

A decade later (1918), and seemingly unaware of Moens' proposals, the prominent German sexologist, Hermann Rohleder, published a thick volume on the problem of hybridization between humans and apes while developing plans for such experiments to take place. These were to be undertaken in an anthropoid station established in 1912 on Tenerife, Canary Islands, by the Prussian Academy of Sciences and a private foundation.

Similarly to Moens, Rohleder thought that the creation of a human-ape hybrid would provide the crucial evidence for evolution. Because of this, he also believed that it was important to use the sperm of a non-European donor (one of the inhabitants of Tenerife with mixed blood) in order to maximize his chances of success. However, Rohleder's plans, like those by Moens, did not progress beyond the conceptual stage. By 1920, the economic situation in Germany made further support of the primate facility on Tenerife impossible and the station was closed (Rossiianov 2002: 292).

Soviet experiments

Only in recently discovered papers originating in Russia, have actual experiments in the field of human-nonhuman combinations been reported. These papers reveal that in the mid-1920s, Joseph Stalin ordered Russia's top animal-breeding scientist, Professor Ilya Ivanov, to turn his skills from horse and animal work to the quest for an ultimate soldier by crossing human beings with apes. According to Moscow newspapers, Stalin told the scientist: 'I want a new invincible human being, insensitive to pain, resistant and indifferent about the quality of food they eat'. The Soviet authorities were indeed struggling at that time to rebuild the Red Army after a series of bruising wars (Rossiianov 2002: 292).

Moreover, the possibility of creating human-nonhuman combinations was also promoted by the Bolshevik officials of the time for ideological reasons. Sergey Novikov, a representative of the Commissariat of Enlightenment (Ministry of Education), referred to the hybridization project as an 'exclusively important problem for Materialism'. In addition, Lev Fridrichson, a representative of the Commissariat of Agriculture, expressed the hope that,

> the topic proposed by Professor Ivanov ... should become a decisive blow to the religious teachings, and may be aptly used in our propaganda and in our struggle for the liberation of working people from the power of the Church. (Ivanov to Lunacharskiy, 17 September 1924; Novikov to Lunacharskiy, 18 September 1924; Friedrichson to Aleksandr Tsyurupa, deputy chairman of Soviet government, 20 September 1924, in Rossiianov 2002: 286)

Similarly, when presenting his proposed research to the Soviet Academy of Sciences in 1925, Prof Ivanov argued that the experiments 'may provide extraordinarily interesting evidence for a better understanding of the problem of the origin of man and of a number of other problems from such fields of study as heredity, embryology, pathology, and comparative psychology' (Rossiianov 2002: 289).

In Ivanov's proposed research, however, no discussion or consideration was ever undertaken relating to the potential ethical difficulties with such experiments (Rossiianov 2002: 289). As Ivanov's negotiations with Western colleagues and patrons indicate, the studies seemed acceptable to both him and his associates because they were taking place in remote colonies, far outside 'civilized' society. From this perspective, the colonies were also considered as 'biological outposts' inhabited by exotic animals and racially inferior people, where the species distance between man and chimpanzee was relatively small (Rossiianov 2002: 290).

Thus, in 1926, the Soviet government and the Academy of Sciences sent Prof Ivanov to French West Africa, backed by a fund of $200,000, to artificially inseminate female chimpanzees with human sperm and to obtain, if possible, a viable hybrid of the two species. This was made possible because Ivanov's African mission was also supported by the directors of the Pasteur Institute in Paris, Emile Roux and Albert Calmette, who allowed Ivanov to work at the institute's primate station in Kindia, French Guinea. On arrival, Ivanov immediately began his studies and eventually inseminated three chimpanzee females with human sperm during the first half of 1927. However, the research failed to produce a hybrid.

At this stage, Ivanov changed tack because of his difficulties in trying to inseminate a resisting chimpanzee. He asked the French authorities whether he could, instead, inseminate native women with the sperm of a dead male chimpanzee in a local hospital. He also insisted that the experiment should be undertaken without the women's knowledge and consent because he was concerned that any woman would, otherwise, resist such an insemination. However, the French authorities refused to give their permission because the studies would be carried out in an established hospital. They suggested, instead, that the procedure could be undertaken 'outside' where no regulations existed. But Ivanov was offended by such a proposal, believing that his research should take place in an objective manner using a clinical and scientific setting (Rossiianov 2002: 299–300).

At the same time, Edwin E. Slosson, director of Science Service, one of the first American non-commercial organisations for the popularization of science, was becoming concerned about religious creationist attacks on Darwinism in the United States. Because of this, he was also looking for instances of hybridization between different species of mammals since he was convinced that the best and most convincing evidence for evolution was to create a new species of higher animals. When he learned about Ivanov's project from Calmette at the Pasteur Institute, he circulated among American newspapers the information that Ivanov would 'try to produce a hybrid between the highest anthropoids and the most primitive of the human race' (Anon 1925; Anon 1926).

It is noticeable that the overtly racist assumptions of this project ('the most primitive of the human race') were shared by Russian, French and American scientists. The fusion of science with explicit racial prejudice had been popularized by the 'eugenic' movement which was very strong in the United States and Europe in the 1920s. This movement continued to flourish until the full revelation of the horrors of Nazi 'science' after the Second World War. Prior to this point there was also much less consciousness of the need for explicit ethical limits on scientific research. The political attraction of these hybrid experiments, as a weapon with which to attack religious creationism, was therefore regarded as sufficient to outweigh any moral scruples about the effects of the research on the experimental subjects.

As a result, Detroit lawyer, amateur biologist and atheism promoter, Howell S. England, who was an acquaintance of Slosson, promised to raise some $100,000 in support of Ivanov's planned experiments. Ivanov even offered to come to the United States for a lecture tour to help raise the money, but his American supporters advised him against such a proposal. The topic of Ivanov's lectures, England warned,

> would be sufficient to raise a perfect storm in our fundamentalist press, all insisting that you be deported and not allowed to land. I would suggest that the best time to have you come to America to lecture would be after the first little anthropoid hybrid shall have been born and ready for exhibition [sic]. We have enough scientists in the United States to assure you after that, not only a safe entrance into the country but a welcome here. (Rossiianov 2002: 295)

In the end, the money was never raised but some interest in hybridization experiments remained amongst American primate researchers.

Meanwhile, the time of Ivanov's mission in West Africa expired in the summer of 1927 without, apparently, undertaking any experiments on women. He left French West Africa with a number of primates which he brought to Sukhum in the southern Soviet Republic of Georgia where the Soviet government established a special primate station. At this station, Ivanov again attempted to arrange further experiments on the artificial insemination of women volunteers, with ideological interests, using the sperm of a 26-year-old orangutan male. However, no evidence exists that these experiments ever took place.

Ivanov's efforts were interrupted when he was arrested by the Soviet secret police in December 1930. For his expensive failure, he was sentenced to five years in jail, which was later commuted to a five year exile in the Central Asian republic of Kazakhstan. Although he was released in 1931, he died in exile in 1932 at the age of 61, without publishing anything about his work either in Africa or in Sukhum. Following Ivanov's death, the details relating to his hybridization program were buried in Soviet archives, and the very fact of his mission was all but forgotten both in Russia and in the West until relatively recently (Rossiianov 2002; Stephen and Hall 2005).

The 'hamster test'

The Hamster Egg Penetration Test (HEPT) was introduced in the United States in the 1970s in order to examine the motility and normality of a man's sperm (Lee and Morgan 2001: 89–90). The test involved mixing golden hamster eggs with human sperm and waiting a few hours, after which a judgement was made relating to the percentage of eggs that were penetrated by the sperm reflecting, in this way, its capacity to fertilize eggs. The importance of the test was its value in studying the chromosomal constitution of human sperm, and hence the male contribution to genetic abnormalities and infertility (which are thought to affect one in 16 of the male population).

At the time when the test was being discussed in the UK by the Warnock Committee in 1984, it was indicated that:

> [b]oth the hamster tests and the possibility of other trans-species fertilisa-tions, carried out either diagnostically or as part of a research project, have caused public concern about the prospect of developing hybrid half-human creatures. (Warnock 1984: 70–1)

Nevertheless, this UK committee then went on to recommend that,

> where trans-species fertilisation is used as part of a recognized programme for alleviating infertility or in the assessment or diagnosis of subfertility it should be subject to licence and that a condition of granting such a licence should be that the development of any resultant hybrid should be terminated at the two cell stage. Any unlicensed use of trans-species ferti-lisation involving human gametes should be a criminal offence. (Warnock 1984: 70–1)

In the subsequent debate over the framing of legislation, though the 1990 legislation addressed the mixing of human sperm with animals from any species and defined an embryo to include an egg in the process of fertilization, some parliamentarians objected to this terminology when a hamster egg was used. They argued that, from a scientific perspective, the term 'fertilization' was not appropriate to the specific situation when human sperm was mixed with a hamster egg, nor was the term 'embryo' appropriate for what resulted. It was noted that clinicians who administered the test rarely if ever described it as 'fertilization'.

However, when other animal species are considered, which are genetically closer to humans, the penetration of an egg by human sperm could raise a number of questions. These included whether, from a scientific perspective, this penetration could be understood as a fertilization event and whether an embryonic organism could be obtained, even if this did not survive for any great length of time.

In the light of these discussions the legislation allowed the mixing of human sperm with the eggs of hamsters, or other animals specified in directions. To assuage the anxiety surrounding the human-nonhuman combinations the law specified that 'anything which forms' was to be destroyed 'not later than the two cell stage' (Human Fertilisation and Embryology Act 1990 Schedule 2, 1(1.f)). This is in contrast to the 14-day limit for experimenting on human embryos.

The Human Fertilisation and Embryology Act 1990 thus permitted the hamster test. However, the subsequent introduction of Intra Cytoplasmic Sperm Injection (ICSI) and other treatments has now made the test effectively obsolete for testing sperm prior to treatment. At the time of writing, the UK Human Fertilisation and Embryology Authority (HFEA) has not given any licences for HEPT to any treatment or research centres since 2003.

It should be noted that while the hamster test was once widely used in fertility treatment and was accepted in legislation, in the debates surrounding the Human Fertilisation and Embryology Bill 2007–2008 the hamster test was frequently cited as a precedent for permitting the creation of human-nonhuman hybrid embryos (Comments of Alan Johnson MP (Hansard 12 May 2008, Column 1068), Phil Willis MP (Hansard 12 May 2008, Column 1119), Evan Harris MP (Hansard 12 May 2008, Column 1139), Dawn Primarolo MP (Hansard 12 May 2008, Column 1159) and Kenneth Clarke MP (Hansard 12 May 2008, Column 1097)). This alleged precedent was given as the reason for permitting in law the mixing of human and nonhuman gametes for a wide range of research purposes, even of closely related species such as chimpanzees, and culturing these embryos for up to 14 days.

This political re-interpretation of the hamster test is a good example of a 'slippery slope' at work in biotechnological regulation. A practice which was initially allowed in a specific setting on the basis that it did not represent a fertilization event and did not lead to the creation of a hybrid embryo was subsequently invoked to justify the deliberate creation of hybrid embryos. This demonstrates that rules which do not incorporate clear lines, but rather admit of exceptions, can sometimes pave the way to much more radical changes in the law.

2

National and international legislation and guidelines

In most countries, no guidelines exist which specifically address human-nonhuman embryonic combinations (Hug 2009). The following awkward question can therefore be asked: how human must a human-nonhuman combination be before human legislation applies or more stringent research rules exist?

Legal definitions

Defining the different kinds of human-nonhuman combination is a major challenge for both scientists and legislators, and one with direct practical consequences. If the categories of human-nonhuman combination are not well described then it will be unclear whether or not an entity is covered by the law relating to human persons. This might mean that it is covered, instead, by different legislation designed to protect animal welfare. Or it might fall outside any law.

In this regard, the source or proportion of the biological material used to create the interspecies entity, in addition to its properties or characteristics, may be crucial in determining whether it comes under human or animal legislation. However, the task of making this judgement should not be underestimated (Eberl and Ballard 2009). It may have fundamental and irreversible implications with respect to the manner in which humanity perceives its identity (Rynning 2009).

From this perspective, there is a pressing need to determine terminology such as the word 'human'. The word can, indeed, be a biological term referring to the species *Homo sapiens* but 'human' can also be understood as an evaluative term that confers a specific status to a being with characteristics that are normally reflected by members of the human species. The evaluative term defining the word 'human' in existing legal texts, moreover, is directly associated with entities that correspond to a biological understanding of 'human'. But in the case of human-nonhuman interspecies embryos, this association is no longer so simple and may pose some fundamental problems in law (Taupitz 2011: 212–13). Thus, defining what is biologically human in the creation of an interspecies combination may be based on a number of criteria such as the source of the biological material used as well as the biological, genetic and 'human functionality' proportions of the resulting entity.

Unfortunately, at the time of writing, in 2012, there is no agreed international legal definition even of the purely human embryo and it is, therefore, unrealistic to expect to find an international consensus on terminology for human-nonhuman entities. Nevertheless, there have been attempts to clarify and standardize terminology and these are surely to be welcomed.

A chimera was originally a Greek mythological fire-breathing female monster with a lion's head, a goat's body and a serpent's tail. In experimental biology, however, terms like chimera and hybrid do not have exact definitions and can sometimes mean quite different things. These terms may even be used interchangeably by molecular biologists when, for example, what is being described is an engineered protein construct (rather than a whole cell or organism). Nevertheless, in a biological context the term 'hybrid' is commonly used for an entity that contains chromosomes from different species (or subspecies etc.) that are present together in the nuclei of most or all its cells. A hybrid in this sense is distinguished from a 'chimera' which contains distinct cell populations each possessing genetic material from just one source. Thus, while there are no universally agreed legal or scientific definitions, there are conventions which are beginning to emerge (as seen, for example, in the Australian Prohibition of Human Cloning Act 2002 and the Canadian Assisted Human Reproduction Act 2004). These are reflected in the following definitions:

   1   A *chimera* is an organism (including an embryo or foetus) that consists of cells of more than one organism.

Human-human embryonic chimeras can occur naturally when non-identical twin embryos fuse in the womb a few days after conception, so that the resulting baby contains genetic material from both embryos. In addition, most twins carry at least a few cells from the sibling with whom they shared a womb, thus chimeras are not uncommon.

It is also possible to define a mosaic which is a biological organism that is made up of more than one genetically distinct population of cells, where these cells are derived from a single fertilised egg. This happens, for example, when a person is made of a combination of cells containing the expected two sex chromosome, such as XY, and some cells in which the Y chromosome is missing. The 'missing Y' cells are created when a Y chromosome is accidentally lost in just some of the cells of a developing embryo.

2   A *human-nonhuman*[1] *chimera* is a biological organism that is made up of genetically distinct populations of human and nonhuman cells. In this context it is possible to define:

   a   *An embryonic or foetal human-nonhuman chimera,* which is:

   - a human embryo or foetus into which at least one cell of a nonhuman life form has been introduced; or

   - a nonhuman embryo or foetus into which at least one cell of a human life form has been introduced.

   b   *A post-natal human-nonhuman chimera,* which is:

   - a human person in which nonhuman animal cells, tissue or organs have been transplanted after his or her birth. For example, experiments have been undertaken, in the past, in which nonhuman hearts have been transplanted into human beings, or

   - a nonhuman animal in which human cells, tissue or organs have been transplanted after it has been born.

Because a chimera is an uneven mixture, a patchwork of different cells, the proportion of different cells can vary. There can be chimeras with very few foreign cells (for example a small valve from a pig's heart transplanted

---

[1]In this definition, and throughout this book, the phrase 'human-nonhuman' combinations has been used in preference to 'human-animal' combinations as though human beings were not animals. While it is true that the word 'animal' is sometimes used to mean 'nonhuman animal', in the present context it is important to recognise that human beings are animals of a particular species. The specific animal nature of human beings grounds both the possibility of human-nonhuman combinations and also the moral evaluation of this possibility.

into a human being) and other chimeras that are much more mixed –
anything up to fifty-fifty (as occurs when an early goat and sheep embryo
are fused to create a 'geep'). The chimeric combinations that are most
likely to raise concern are those which involve human brain or reproductive
tissue. These more disturbing possibilities are sometimes distinguished in
legislation or policy guidelines.

3   A *human-nonhuman hybrid* is a biological organism in which most or
all cells have combined human and nonhuman origin. These include:

   a   A *true hybrid,* whereby:

   ● a human ovum has been fertilized by sperm of a nonhuman
   life form.

   ● an ovum of a nonhuman life form has been fertilized by human
   sperm.

In this case, the hybrid will be approximately 50 per cent human and 50
per cent nonhuman. A combination created by the fusion of gametes from
two different species is the central case of an interspecies hybrid and
sometimes the word 'hybrid' is used with this restricted use. However,
because the word is often applied to a wider range of human-nonhuman
combinations, the hybrid created by a fusion of gametes is sometimes
referred to as a *true hybrid* (for example HFEA 2007a, p. 9). There is
some irony in this term for a hybrid is not 'true' to any one species.
The Japanese legislation refers to true hybrid embryos as amphimictic
(meaning 'mixed from both' – by fertilisation), but this terminology is not
common.

   b   A *cybrid,* whereby:

   ● the nucleus of a cell of a nonhuman life form has been
   introduced into a human ovum stripped of its own
   chromosomes.

   ● the nucleus of a human cell has been introduced into an ovum
   of a nonhuman life form stripped of its own chromosomes.

As well as the fusion of gametes, it is now possible to create a new
individual by removing the nucleus of an egg cell and replacing it with
the nucleus of an ordinary body cell from an adult. This is what was
famously done in the case of Dolly the sheep. Because of the way she was
produced, Dolly was (almost) genetically identical to her parent. The kind of
technique is often termed 'cloning'. The clone is identical to a pre-existing
original. This idea has been exploited in many works of science fiction,

of which Ira Levin's 1976 novel *The Boys from Brazil* is a good example. Nevertheless, the term 'clone' has been criticised as potentially misleading e.g. UNESCO 2009, though it is noticeable that, even though the UNESCO committee disapproved of the term, they felt compelled to use the word 'cloning' in the title of the report – which attests to the extent to which the term has become common parlance. Scientists generally prefer the term Somatic Cell Nuclear Transfer (SCNT). Interspecies SCNT leads to an entity that has over 99% of its DNA from one species. In this context, it is noteworthy that South Korea, which prohibited human-nonhuman SCNT in 2008 referred to the entities created as 'interspecies clones'. This kind of combination is certainly much closer to a clone than it is to a true hybrid. Nevertheless, the emerging standard terminology is to refer to these entities as cytoplasmic hybrids or 'cybrids'. This is the terminology used in this book.

    c   *Other kinds of hybrids,* in which there is a human ovum or an ovum of a nonhuman life form that otherwise contains chromosomes from both a human being and a nonhuman life form.

In addition to true hybrids and cybrids, there may be various other ways of creating an entity that combines human and nonhuman chromosomes. For example, chromosomes of one species might be added to the egg of another species before fertilisation.

It is difficult to know what techniques will emerge in the future. Many legal systems attempt to have 'catch all' phrases to cope with future developments but these are very difficult to formulate. Nevertheless, in terms of the categories accepted in this book, an ovum that contains chromosomes of two different species is a hybrid.

In short, what hybrids have in common is that they blend genetic information from different lines or even species, at a cellular level. They do not present a 'patchwork' appearance, and are not always less fit than their progenitors (Animal Procedures Committee 2001, pp.18–20).

4   A *human-nonhuman transgenic individual* is a human being into whom one or a small number of nonhuman genes have been transferred, or a nonhuman life form into which one or a small number of human genes have been transferred.

It would seem that the definition of hybrid given above should in fact cover transgenic individuals. They also blend elements from different species at the cellular and sub-cellular level. If cybrids, in which over 99% of the DNA is from one species, are commonly described as a kind of hybrid then individuals with a small number of genes from a different species incorporated into their DNA should also count as hybrids. For this reason the word 'hybrid' is sometimes used to cover transgenic individuals. Nevertheless, gene transfer has long been understood to be a distinct category from hybridisation and the most common usage does not include transgenic entities as 'hybrids'. This constitutes an exception to the definition we have given of hybrid. In the interests of consistency it would probably be better if transgenic individuals were regarded as a kind of hybrid or, alternatively, if the term hybrid was confined to true hybrids. Nevertheless, in the interests of supporting a consensus on terminology, where this seems to be emerging, this book will follow the convention of using 'hybrid' to cover cybrids but not to cover transgenic individuals.

5 A *human-nonhuman combination* is a biological organism which combines human and nonhuman material.

In this book the term 'human-nonhuman combination' is used as the over-arching category for human-nonhuman chimeras, hybrids and transgenic individuals. There is little international legal or scientific agreement as to a common term that can encompass all three categories. Occasionally the term 'chimera' is used in this over-arching sense, or again the term 'hybrid' may be given this broader definition. The disadvantage of using either 'chimera' or 'hybrid' as the over-arching term is that this obscures the differences between chimeras and hybrids. The adjective 'interspecies' is a better candidate for an over-arching term, as in the phrase 'interspecies embryos'. Nevertheless this term is too broad for the primary purposes of legislation – which is generally concerned not with combinations of two different nonhuman species but specifically with human-nonhuman combinations.

In the first draft of United Kingdom legislation in this area, the *Human Fertilisation and Embryology Bill 2007–2008*, the term 'interspecies embryo' was used as the overarching category (see below). However, during the passage of the bill in parliament this phrase was criticised and the term 'human admixed embryo' was used instead. This has the advantage of clarifying that the combination is predominantly human and also clarifies that the entities in question are embryos (not merely 'cells' or 'biological material'). However, the definition is rather coy about the nature of the admixing. Admixed with what? Admixed how? Taken on face value the term would seem to cover, for example, human chimeric embryos (created from admixing two different human embryos). What is missing from the term 'human admixed' is clarification that the admixing is specifically with a nonhuman animal. As there is very little agreement internationally about an overarching term, this book uses and recommends 'human-nonhuman combination'.

International legislation

As the creation for research of human-nonhuman interspecies embryos and foetuses raises important scientific questions, such combinations also challenge the very important ethical, social and legal understanding of what it is to be a member of the human species (Taupitz 2011: 211). But on account of the novelty of human-nonhuman combinations, they have rarely been considered on the international stage and no binding legal texts exist. Most international conventions only address either human beings or nonhuman animals and have not yet examined entities which might not fit easily into either category. This is unfortunate, since a real legal vacuum exists with the potential that future human-nonhuman experiments may not be appropriately regulated. Moreover, because this field of science does not demand huge investments, it may encourage funders and researchers to by-pass any strict provisions in their national legislation by travelling to countries where regulations are far more liberal. An international discussion and consultation is, therefore, urgently required in this field.

The little that does exist on the international level includes the following.

United Nations

Universal Declaration of Human Rights (Adopted and proclaimed by United Nations General Assembly resolution 217 A (III) of 10 December 1948).

Article 1 indicates that:

> All human beings are born free and equal in dignity and rights. They are endowed with reason and conscience and should act towards one another in a spirit of brotherhood.

This is a key article which has been incorporated into many international conventions and into the national legislation of many countries. However, neither this article nor the many laws and conventions that are based upon it define what is meant by 'human being'. This, it should be recalled, was the focus of the trial in the novel *Borderline*. If a true human-nonhuman hybrid were ever to be born, the most likely being a humanzee, then the question would immediately arise: Would such a creature be born 'free and equal [to human beings] in dignity and rights'?

Council of Europe (47 Member States)

Committee of Ministers of the Council of Europe: European Convention on Human Rights and Biomedicine (ETS No. 164)

From the completely human side of the spectrum this Convention on Biomedicine is one of the most important legally binding instruments in existence in Europe (for the countries that have ratified it) and shapes legislation in many countries. It entered into force on the 1st of December 1999. The European Convention has not been ratified by all members of the Council of Europe, notable exceptions being the United Kingdom (which has a more permissive national biopolicy) and Germany (which has a more restrictive national biopolicy). Nevertheless, with 29 ratifications and 6 further signatories the Convention (as of January 2012) is an important international instrument.

Article 1 (Purpose and object) indicates that:

> Parties to this Convention shall protect the dignity and identity of all human beings and guarantee everyone, without discrimination, respect for their integrity and other rights and fundamental freedoms with regard to the application of biology and medicine.

Article 11 (Non-discrimination) indicates that:

> Any form of discrimination against a person on grounds of his or her genetic heritage is prohibited.

Article 13 (Interventions on the human genome) indicates that:

> An intervention seeking to modify the human genome may only be undertaken for preventive, diagnostic or therapeutic purposes and only if its aim is not to introduce any modification in the genome of any descendants.

Article 18 (Research on [human] embryos *in vitro*) indicates that:

> 1. Where the law allows research on [human] embryos *in vitro*, it shall ensure adequate protection of the embryo.

2. The creation of human embryos for research purposes is prohibited.

As with the United Nations Declaration, the European Convention guarantees rights to human beings (Article 1) or persons (Article 11) but does not define these terms or set out criteria for humanness or personhood. It seems to be assumed that no reasonable confusion is possible as to whether a being counts as a human being but in the future this may no longer hold. Similarly in Article 13 the phrase 'human' genome is not defined. What human percentage does a genome have to be to count as human?

It is already the case that questions arise in the application of Article 18. What counts as a 'human embryo'? If an embryo is created from the egg of a cow or a rabbit but with the cell nucleus from a human being, is that a human embryo? Is the creation of such an embryo prohibited by the European Convention? Note that the Convention does not forbid research on human embryos but it forbids the *creation* of human embryos for research. This distinction between using and creating for use is related to the idea of instrumentalization prominent in German speaking discussion of bioethics and influenced by the philosophy of Immanuel Kant. This idea does not imply or presuppose that the human embryo is a 'human person', but it does imply that the human embryo shares in human nature to such an extent that it should not be created as a means to an end. However, if the basis of protection is sharing a common human nature, the question arises as to *how much* of a share is needed. Only if admixed human hybrids are human enough (as it were) to count as human embryos will their creation fall foul of the Convention (Council of Europe 1997, see Article 18.2 quoted above).

Such embryos have already been created in countries which have not signed the European Convention on Human Rights and Biomedicine (such as China, the United States and the United Kingdom). It has yet to be clarified whether they may lawfully be created by countries which have ratified this Convention.

Interestingly, if an embryo is created by using the nuclear gene set of a human individual and an animal egg, stripped of its chromosomes, and this embryo is itself considered as 'nonhuman', then the Additional Protocol to the above Convention on the Prohibition of Cloning Human Beings would still prohibit the creation of such an embryo if this would give rise to a human being (though this would be dependent on the manner in which the Additional Protocol interacts with the Convention on Human Rights and Biomedicine). In this context it should be noted that the percentage of nonhuman material may decrease with development. Proportionally it could become more human.

The Additional Protocol states (Council of Europe 1998a, Article 1):

1. Any intervention seeking to create a human being genetically identical to another human being, whether living or dead, is prohibited.

2. For the purpose of this article, the term human being 'genetically identical' to another human being means a human being sharing with another the same nuclear gene set.

In this regard, the term 'human being' was not defined in the Protocol since it was left to domestic law to clarify. For example, the Netherlands has declared that it interprets the term 'human being' as referring exclusively to a human being who has been born.

Committee of Ministers of the Council of Europe: European Convention for the Protection of Vertebrate Animals used for Experimental and other Scientific Purposes CETS No.: 123

From the completely animal side of the spectrum this Convention on the Protection of Vertebrate Animals has become central in Europe with the European Union having ratified it in its entirety. It entered into force on 1 January 1991. According to Article 1:

Section 1:   This Convention applies to any animal used or intended for use in any experimental or other scientific procedure where that procedure may cause pain, suffering, distress or lasting harm. It does not apply to any non-experimental agricultural or clinical veterinary practice.

Section 2.a: In this Convention ... 'animal', unless otherwise qualified, means any live non-human vertebrate, including free-living and/or reproducing larval forms, but excluding other foetal or embryonic forms;

Section 2.c: In this Convention ... 'procedure' means any experimental or other scientific use of an animal which may cause it pain, suffering, distress or lasting harm, including any course of action intended to, or liable to, result in the birth of an animal in any such conditions, but excluding the least painful methods accepted in modern practice (that is 'humane' methods) of killing or marking an animal.

Article 7:   When a procedure has to be performed, the choice of species shall be carefully considered and, where required, be explained to the responsible authority; in a choice between procedures, those should be selected which use the minimum number of animals, cause the least pain, suffering, distress or lasting harm and which are most likely to provide satisfactory results.

As an addition to this Convention, the following legal instrument was prepared: Protocol of Amendment to the European Convention for the Protection of Vertebrate Animals used for Experimental and other Scientific Purposes CETS No.: 170 (Council of Europe 1998b). Since its entry into force on the 2nd of December 2005, this Protocol is an integrant part of the convention (Council of Europe 1986a) and is no longer open for signature or ratification. 19 European countries have ratified this Convention in addition to the European Union as a whole.

Committee of Ministers of the Council of Europe: European Convention for the Protection of Pet Animals (ETS No. 125)

Another European Convention that may be relevant to the protection of nonhuman animals is the one addressing pet animals (Council of Europe 1987) which came into force on the first of May 1992. This has been ratified by 22 European Countries.

Article 1 (Definitions) it is notified that:

1. By pet animal is meant any animal kept or intended to be kept by man in particular in his household for private enjoyment and companionship.

Article 3 (Basic principles for animal welfare) indicates that:

1. Nobody shall cause a pet animal unnecessary pain, suffering or distress.
2. Nobody shall abandon a pet animal.

Parliamentary Assembly of the Council of Europe, Recommendation 1046 (1986) on the use of human embryos and foetuses for diagnostic, therapeutic, scientific, industrial and commercial purposes

As already indicated, with respect to human-nonhuman interspecies entities, no guidelines or regulations are really in existence in the Council of Europe apart from a non-legally binding Recommendation on Xenotransplantation. The only text in existence that may relate to the topic comes from the Parliamentary Assembly.

The Parliamentary Assembly of the Council of Europe does not prepare legally binding legislation such as the Conventions of the Committee of Ministers. However, it can prepare what is called 'soft' law which may be used, for example, by the judges of European Court of Human Rights to advise and inform their decisions. One such text is 'Recommendation 1046 (1986) on the use of human embryos and foetuses for diagnostic, therapeutic, scientific, industrial and commercial purposes' (Council of Europe 1986b), which addresses some aspects of human-nonhuman research.

In Article 14 (A) it recommends that the Committee of Ministers call on the governments of the member states:

iii. to forbid any creation of human embryos by fertilisation *in vitro* for the purposes of research during their life or after death;

iv. to forbid anything that could be considered as undesirable use or deviations of these techniques, including: …

the implantation of a human embryo in the uterus of another animal or the reverse;

the fusion of human gametes with those of another animal (the hamster test for the study of male fertility could be regarded as an exception, under strict regulation);

the fusion of embryos or any other operation which might produce chimeras.

Parliamentary Assembly of the Council of Europe: Embryonic, Foetal and Post-natal Animal-Human Mixtures, Doc. 10716, 11 October 2005, Motion for a resolution presented by Mr Wodarg and others of the committee on Culture, Science and Education.

It is in the light of this dearth of international provision on human-nonhuman combinations that the Council of Europe considered a motion on the issues raised by human-nonhuman combinations.

In this draft document it is suggested that the Parliamentary Assembly of the Council of Europe should invite the governments of the member states to initiate an extensive consultation and reflection on the issues surrounding the creation of human-nonhuman combinations. The document also encourages the Assembly to evaluate the scientific facts, general ethical arguments, different risks and social consequences and, if necessary, to table its own recommendations.

The operation of the Council is quite cumbersome. The document is a motion presented by a member of one committee. This committee does not have the power to make proposals for amendments to Conventions. Rather the document recommends to the Committee of Ministers to entrust the Steering Committee on Bioethics of the Council of Europe to address the ethical issues arising from the creation of human-nonhuman combinations and to insert this task into the ongoing work on additional protocols to the European Convention on Human Rights and Biomedicine. Though the document was produced in 2005 there has been no further activity from the Council of Europe on the issue.

European Union (27 member states)

From a European Union perspective, it is also the case that no legal instruments directly relating to human-nonhuman entities are in existence. However, a recent new legal instrument has been prepared in the form of Directive (2010/63/EU) on the protection of animals used for scientific purposes which has become European Law in the EU on the 22 September 2010. This now replaces the older Directive (86/609/EEC) of the 24 November 1986 on the approximation of laws, regulations and administrative provisions of the Member States regarding the protection of animals used for experimental and other scientific purposes.

Directive (2010/63/EU) on the protection of animals used for scientific purposes

This directive adds to its older version regarding the protection of animals by introducing measures aiming to improve the wellbeing of animals used for experiments while allowing vital research to continue. It also seeks to integrate and expand the previously mentioned Council of Europe Convention for the protection of vertebrate animals used for experimental and other scientific purposes while enabling member states to ensure stricter national measures. It emphasizes that 'Animals have an intrinsic value which must be respected'.

Article 1 (Subject matter and scope) it is indicated:

Section 3: This Directive shall apply to the following animals:

(a) live non-human vertebrate animals, including:

(i) independently feeding larval forms; and
(ii) foetal forms of mammals as from the last third of their normal development;

(b) live cephalopods.

Section 4: This directive shall apply to animals used in procedures, which are at an earlier stage of development than that referred to in point (a) of paragraph 3, if the animal is to be allowed to live beyond that stage of development and, as a result of the procedures performed, is likely to experience pain, suffering, distress or lasting harm after it has reached that stage of development.

The directive also emphasizes in Article 8 (Non-human primates) that the use of nonhuman primates should only be possible in very restricted conditions since they are the animals with the most genetic proximity to human beings as well as having highly developed social and behavioural skills. They are only to be used in biomedical areas where they are essential for the benefit of human beings, for which no other alternative replacement methods are yet available. This included research into life-threatening and debilitating conditions endangering human beings.

The directive also states the different requirements which are necessary to perform the research, including different forms of authorisation by a competent authority for (1) the breeders, suppliers and users of animals, and (2) the specific projects being considered.

In this regard, Article 13 (Choice of methods) states that:

Section 2

In choosing between procedures, those which to the greatest extent meet the following requirements shall be selected:

(a) use the minimum number of animals;
(b) involve animals with the lowest capacity to experience pain, suffering, distress or lasting harm;
(c) cause the least pain, suffering, distress or lasting harm;

and are most likely to provide satisfactory results.

There is a clear contrast between this directive and the international instruments we have considered hitherto. The directive gives a detailed definition of 'animal' which, for example, exclude embryos, early foetuses, and human beings, and which includes all other vertebrates. However, this definition does not address the issue of human-nonhuman combinations. In relation to human-nonhuman combinations the definition 'live non-human vertebrate' begs the question. What is nonhuman? Might a humanzee qualify as an experimental animal? Or should he or she be protected as a human subject?

National legislation

Because the time necessary to prepare an international instrument is considerable, it is not surprising that it has been at a national level that the question of human-nonhuman combinations has thus far been addressed.

In this regard, it was in countries, such as Argentina and Germany that first passed laws in relation to human-nonhuman combinations.

Argentina

Province of Neuquén Law n° 2258 on the creation of a permanent provincial commission on fecundation and genetic research of 15 October 1998

This statute is one of a number that is referred to in the Report of the International Bioethics Committee on Human Cloning and International Governance (UNESCO 2009). The issue of human-nonhuman combinations is distinct from the question of human cloning but they are related and often fall under the same national legislation. In the context of prohibiting human cloning, trade in human gametes and the cryopreservation of human embryos, the Argentinian law forbids the creation of hybrid embryos.

Germany

The Embryo Protection Act [Embryonenschutzgesetz] of 13 December 1990

In Germany, the creation of human-nonhuman interspecies embryos was already envisaged in 1990 with the Embryo Protection Act. In this legal instrument, Section 7 (Creation of chimeras and hybrids) states that whosoever attempts to:

1 Combine embryos with different genetic information, through the use of at least one human embryo, into a united cell structure, or

2 Combine a human embryo with a cell that contains genetic information different from the embryo's cells and induces it to further develop, or

3 Generate an embryo capable of development through the fertilisation of a human egg cell with animal sperm or through the fertilisation of an animal egg cell with human sperm, or

4 Transfer an embryo obtained through a procedure defined in (1), (2) and (3) above into a woman or an animal, or

5 Transfer a human embryo into an animal.

will be liable to a sentence of up to five years imprisonment or a fine.

Moreover, it should be indicated that Section 2 (1) of the German Embryo Protection Act prohibits the use of human embryos and totipotent cells for research.

Japan

The Law Concerning Regulation to Human Cloning Techniques and Other Similar Techniques (30 November 2000)

While the legal provisions of Argentina and Germany concern the creation of human-nonhuman combinations, the law enacted by Japan in 2000 prohibits only the transfer of such an embryo into a woman or a nonhuman animal. This leaves open the possibility of creating hybrid or chimeric embryos *in vitro* for research purposes unless a temporary moratorium is introduced by the government to prohibit such as procedure.

Article 3 of the Japanese law indicates that no person shall transfer a human somatic cloned embryo, a human-animal 'amphimictic' embryo (that is, a true hybrid), a human-animal (cytoplasmic) hybrid embryo or a human-animal chimeric embryo into the uterus of a human or an animal (Roetz, 2006: 4).

Given this precision it is not surprising that Japan was also the first country to prohibit the implantation of an embryo formed by means of induced pluripotent stem cells (iPS cells). These are cells taken from an adult which have been modified biochemically so that they behave in a similar manner to embryonic cells. The creation of human iPS cells was achieved for the first time in mice in 2006 and in human beings in 2007. Already in February 2008, Japan's science ministry sent all universities and research agencies a notification specifically forbidding,

> the implantation of an embryo made with iPS cells into human or animal wombs, the production of an individual in any other way from iPS cells, the introduction of iPS cells into an embryo or fetus, and the production of germ cells from iPS cells. (Cyranoski 2008)

Estonia

Penal code §130, adopted on 6 June 2001

In 2001, Estonia passed a law which followed the Argentinean and German rather than the Japanese model, i.e. banning the creation of hybrid embryos. In relation to terminology it was an advance on the Argentinean law in that it expressly covered both 'hybrids' and 'chimeras'.

Switzerland

Federal Law on Embryonic Stem Cell Research of 19 December 2003, came into force on 1 March 2005

The law in Switzerland (Article 3.c) expressly forbids not only the creation of clones, chimeras, and hybrids but also the use of cloned, chimeric, or hybrid embryos to produce stem cells ('de produire des cellules souches embryonnaires à partir d'un clone, d'une chimère ou d'un hybride, ou d'utiliser de telles cellules').

In this regard, Article 2 of the Federal Reproductive Medicine Act of 18 December 1998 (Schweizer and Bernhard 2009: 161) defines:

- *cloning* as the artificial creation of genetically identical species,

- a *chimera* as the combination of totipotent cells from two or more genetically different embryos into a united cell structure,

- a *hybrid* as the introduction of a nonhuman sperm into a human ovum or of a human sperm into a non-human ovum,

- *totipotent cells* as embryonic cells that are still capable of developing into the most divers tissues.

Again it represents the more rigorous restrictions of Argentina, Germany and Estonia rather than the partly permissive stance embodied in the Japanese law. However, it is on a similar level of detail to the Japanese legislation.

The report of the UK Human Fertilisation and Embryology Authority in 2007 was therefore quite wide of the mark when it declared that 'To date only Australia, Canada and the USA have passed legislation on human-animal embryos' (HFEA 2007b, 3.2). The Canadian legislation had in fact been anticipated by Argentina, Germany, Japan, Estonia and Switzerland and would soon be followed by significant legislation in South Korea. Nevertheless, the Canadian legislation is worth highlighting as it clarified some of the terminology in this area and helped shape legislation in other countries.

Canada

Assisted Human Reproduction Act 2004

The Canadian legislation is very helpful since it begins by carefully defining the different terms which are later used by other jurisdictions.

Clause 3 in this Act defines:

An 'embryo' as:

> a human organism during the first 56 days of its development following fertilization or creation, excluding any time during which its development has been suspended, and includes any cell derived from such an organism that is used for the purpose of creating a human being.

A 'chimera' as:

> (a) an embryo into which a cell of any non-human life form has been introduced; or
> (b) an embryo that consists of cells of more than one embryo, foetus or human being.

A 'hybrid' as:

> (a) a human ovum that has been fertilized by a sperm of a non-human life form;

(b) an ovum of a non-human life form that has been fertilized by a human sperm;

(c) a human ovum into which the nucleus of a cell of a non-human life form has been introduced;

(d) an ovum of a non-human life form into which the nucleus of a human cell has been introduced; or

(e) a human ovum or an ovum of a non-human life form that otherwise contains haploid sets of chromosomes from both a human being and a non-human life form.

The Canadian legislation then continues by describing what is prohibited. Clause 5 indicates that it is prohibited to:

Create a chimera, or transplant a chimera into either a human being or a nonhuman life form; or

Create a hybrid for the purpose of reproduction, or transplant a hybrid into either a human being or a nonhuman life form.

It is noticeable and commendable that the Canadian law does not use the term 'animal' to refer to nonhuman animals, but rather uses the phrase 'non-human life form'. It is also commendable that the law carefully defines what it means by chimera and hybrid. The Canadian legislation has had an important role in the move towards established legal categories and agreed terminology.

Research involving the transplantation of human stem cells into nonhuman embryonic, foetal or adult animals is not prohibited in law. However, the Canadian Institutes of Health Research (CIHR) has issued guidelines that apply to all Canadian researchers and research institutions that receive funding from CIHR and other federal funding agencies. The 2010 guidelines (CIHR 2010, Paragraph 8.2) expressly prohibit the creation of animal chimera embryos and foetuses for research which includes:

- Research in which human or non-human pluripotent cells are combined with a human embryo.

- Research in which human or non-human pluripotent cells are grafted to a human foetus.

- Research in which human pluripotent cells are combined with a non-human embryo.

- Research in which human pluripotent cells are grafted to a non-human foetus.

South Korea

Bioethics and Biosafety Act (1 January 2005)

In South Korean legislation, Article 12 (Prohibition on the Transfer of Embryos between Two Different Species) indicates that:

1 No person shall implant a human embryo into the uterus of an animal; nor shall anyone implant an animal embryo into a human uterus.

2 No person shall perform any of the following acts:

● Fertilising a human oocyte with an animal sperm, or vice versa, for any purpose other than that of testing human sperm cells;

● Implanting an animal's somatic cell nucleus into a human oocyte whose nucleus has been removed;

● Fusing a human embryo with an animal embryo; or

● Fusing a human embryo with another embryo of non-identical genetic information.

3 No person shall transfer the products of any of the acts described in Article 12 (2) into the uterus of a human being or animal.

It is widely agreed that at the turn of the millennium South Korea had a permissive legislative framework and public policy environment which favoured biotechnical innovation over bioethical restraint (Harmon and Kim 2008). This was the background to the actions of Dr Hwang Woo-Suk who claimed in 2004 to have created the first cloned human embryo. It gradually emerged that he had obtained the material for his research by paying women to donate their eggs and also by obtaining eggs from his female research assistants. This raised significant ethical concerns. Later it was revealed that his results were in fact falsified and his claim to have cloned a human embryo was fraudulent (Saunders and Savulescu 2008). This scandal had not yet arisen when the Bioethics and Biosafety Act 2005 was being drafted. Nevertheless the act did represent a modest move away from unrestrained biotech freedom and towards ethical governance.

In May 2008, South Korea's parliament passed a further law prohibiting the implanting of a human somatic cell nucleus into a nonhuman egg. This procedure was described as 'cross-species cloning' (APF 2008) and both this description and the prohibition were clearly shaped by the Hwang cloning scandal. Interestingly, the law was passed in South Korea in the same month that parliamentarians in the United Kingdom voted expressly to permit the same procedure. In the United Kingdom this particular form of human-nonhuman combination was not termed as a 'cross-species clone' but as a 'cytoplasmic hybrid' or 'cybrid'. The terminological differences seem in part to reflect ethical differences. 'Clone' has become a pejorative term and is generally avoided by those who wish to permit SCNT to create embryos for research purposes only.

Australia

Prohibition of Human Cloning for Reproduction Act 2002

In Section 8 of the Australian legislation it is indicated that:

> *chimeric embryo* means:
>
> (a) a human embryo into which a cell, or any component part of a cell, of an animal has been introduced; or
> (b) a thing declared by the regulations to be a chimeric embryo.

Prohibition of Human Cloning for Reproduction and the Regulation of Human Embryo Research Amendment Act 2006

This 2006 Act amends the Prohibition of Human Cloning Act 2002 and the Research Involving Human Embryos Act 2002.

Schedule 2 (2–3) indicates that:

> *human embryo* means a discrete entity that has arisen from either:
>
> (a) the first mitotic division when fertilisation of a human oocyte by a human sperm is complete; or
> (b) any other process that initiates organised development of a biological entity with a human nuclear genome or altered human nuclear genome that has the potential to develop up to, or beyond, the stage at which the primitive streak appears;
>
> and has not yet reached 8 weeks of development since the first mitotic division.

> *hybrid embryo* means:

(a) an embryo created by the fertilisation of a human egg by animal sperm; or

(b) an embryo created by the fertilisation of an animal egg by human sperm; or

(c) a human egg into which the nucleus of an animal cell has been introduced; or

(d) an animal egg into which the nucleus of a human cell has been introduced; or

(e) a thing declared by the regulations to be a hybrid embryo.

In this 2006 Act, it is prohibited to:

- Place a human embryo clone in the body of a human or the body of an animal.

- Create a human-nonhuman chimeric embryo.

- Develop a hybrid embryo for a period of more than 14 days, excluding any period when development is suspended.

- Create a human-nonhuman hybrid embryo without a licence.

A licence to create or develop a hybrid embryo can only be issued for the purposes of testing sperm quality through the fertilisation of an animal egg by a human sperm in an accredited Assisted Reproduction Technology centre – up to, but not including, the first mitotic division. Note also that under the Prohibition of Human Cloning for Reproduction and the Regulation of Human Embryo Research Amendment Act 2006, Schedule 2, Item 15, Subsection 20(1), a licence cannot be obtained to create a hybrid embryo by introducing the nucleus of a human cell into an animal egg.

In Australian legislation, it is also prohibited to:

- Place a human embryo in an animal.

- Place an animal embryo in the body of a human for any period of gestation.

Note here that the definitions are closely parallel to those of the Canadian law which clearly influenced the legislators. It is significant that the continued prohibition of hybrid and chimeric embryos was agreed in the context of a liberalisation of the law on cloning human embryos for research. The Australian law permitted the use of SCNT to create human embryos for research purposes but did not generally permit the creation of human-nonhuman hybrid, cybrid or chimeric embryos. This was a step too far. The exception for 'testing sperm quality' relates to the 'hamster test' and echoes the provisions in the United Kingdom's Human Fertilisation and Embryology Act 1990.

France

One of the most recent legislations to address human-nonhuman embryonic combinations was enacted in France in 2011. This benefitted somewhat from the discussions that had taken place in the preparation of similar legislation in the UK in 2008 and the subsequent scientific developments.

In this new Law, n° 2011–814 of 7 July 2011 relating to bioethics, Article 40, modified the Public Health Code, art. L2151-2 (V) to indicate that the *in vitro* conception of an embryo or the constitution of a human embryo through cloning for research purposes is prohibited. It also banned the creation of transgenic or chimeric embryos.

This means that the legislators in France did not consider any liberalization of the law on the creation of human-nonhuman embryos as being either appropriate or suitable.

United States

As of 2009 there is no national legislation in the United States to parallel the laws of Australia and Canada. In most states, almost all forms of research would be possible, from a legal perspective, if it is privately funded and is not intended to produce a commercial product (Stoltzfus Jost 2009: 776)

On 9 March 2009, President Barack Obama issued Executive Order 13505 entitled 'Removing Barriers to Responsible Scientific Research Involving Human Stem Cells' (NIH 2009). In this decision it is indicated that the following research is ineligible for National Institute of Health funding:

- Research in which human pluripotent stem cells[2] are introduced into non-human primate blastocysts.

- Research involving the breeding of animals where the introduction of pluripotent stem cells may contribute to the germ line.

From a legislative perspective, there have been a number of attempts, since 2005, to bring in legal instruments and the drafts of this law are worth considering.

---

[2] Human pluripotent stem cells are human cells that are capable of dividing without differentiating for a prolonged period in culture, and are known to develop into cells and tissues of the three primary germ layers of an embryo.

US Draft Human Chimera Prohibition Act of 2005 (S.1373)

This was introduced in the Senate of the United States by Mr Brownback (R-Kan) with four co-sponsors.

Section 301. Definitions
In this chapter the following definitions apply:

(1) HUMAN CHIMERA – The term 'human chimera' means –

(A) a human embryo into which a non-human cell or cells (or the component parts thereof) have been introduced to render its membership in the species Homo sapiens uncertain through germline or other changes;
(B) a hybrid human/animal embryo produced by fertilizing a human egg with non-human sperm;
(C) a hybrid human/animal embryo produced by fertilizing a non-human egg with human sperm;
(D) an embryo produced by introducing a non-human nucleus into a human egg;
(E) an embryo produced by introducing a human nucleus into a non-human egg;
(F) an embryo containing haploid sets of chromosomes from both a human and a non-human life form;
(G) a non-human life form engineered such that human gametes develop within the body of a non-human life form; or
(H) a non-human life form engineered such that it contains a human brain or a brain derived wholly or predominantly from human neural tissues.

(2) HUMAN EMBRYO – The term 'human embryo' means an organism of the species Homo sapiens during the earliest stages of development, from 1 cell up to 8 weeks.

Sec. 302. Prohibition on human chimeras.

(a) In General – It shall be unlawful for any person to knowingly, in or otherwise affecting interstate commerce –

(1) create or attempt to create a human chimera;
(2) transfer or attempt to transfer a human embryo into a non-human womb;

(3) transfer or attempt to transfer a non-human embryo into a human womb; or

(4) transport or receive for any purpose a human chimera.

The terminology of this draft legislation did not yet reflect the distinction between 'hybrid' and 'chimera' seen in the Canadian and Australian law. As terminological clarity is important here, this was a weakness and one corrected, to some extent, in later draft bills. In other ways, the legislation is commendably detailed and comprehensive in its description of different categories of human-nonhuman combinations. The explicit mention of brain or neural tissues was also helpful in drawing attention to a subcategory of chimera that intuitively seems more troubling.

To the disappointment of its sponsors, this bill, like most bills and resolutions that enter the legislative process, did not get past the committee stage.

US Draft Human-Animal Hybrid Prohibition Act of 2007 (S. 2358)

In November 2007, Senators Sam Brownback (R-Kan) and Mary Landrieu (D-La) re-introduced the bill as the Draft Human-Animal Hybrid Prohibition Act of 2007 (S. 2358). It attracted 18 co-sponsors.

This Draft was identical to the Draft Human Chimera Prohibition Act 2005 except that the phrase 'human chimera' was replaced by 'human-animal hybrid' and in 1 (A) the phrase 'through germline or other changes' was deleted. The term 'human-animal hybrid' was an improvement on 'human chimera' as a generic term for human-nonhuman combinations, but, as argued above, it would have been better not to use either 'chimera' or 'hybrid' for the overall term but to define chimeras and hybrids separately, as in the Canadian legislation. The term 'human-animal hybrid' also represents a use of animal to mean nonhuman animal. This is unhelpful and is avoided by the Canadian legislation.

By 2007 the Republicans had lost control of the Senate and it is not surprising that this, predominantly Republican bill (only one of its co-sponsors was Democrat) failed to make progress on this its second attempt.

US Draft Human-Animal Hybrid Prohibition Act of 2008 (H.R. 5910)

A similar Draft Bill was introduced to the US Congress, in April 2008, by Rep Christopher Smith (R-N.J.) with 13 co-sponsors (all Republican) as the Human-Animal Hybrid Prohibition Act of 2008 (H.R. 5910). This also failed to make progress.

US draft Human-Animal Hybrid Prohibition Act of 2009 (S. 1435)

This version of the bill, identical to the Draft Human-Animal Hybrid Prohibition Act of 2007 (S. 2358) was introduced in July 2009. The increasing number of co-sponsors for this Bill may show a development in support for this legislation within the Republican Party, but of these co-sponsors still only one was Democrat, Senator Mary Landrieu of Louisiana. The same partisan pattern was also reflected in the sponsors of the bill introduced into Congress in 2008.

The failure to garner greater Democrat support relates in part to the conflation of this issue with the embryonic stem cell issue, to which it is only indirectly related. After the election of 2008, any appearance of restraining biotechnology was a difficult sell in the Democratic Party. It might also have been easier to attract support had the legislation been less ambitious. Had Senator Brownback confined the prohibition to the 'bright line' recommended by the President's Council on Bioethics in 2004 and prohibited only 'true hybrids', this might have provided a better basis for a bipartisan consensus.

Louisiana Senate Bill No. 115 (Act 108)

The attempts of Senator Brownback and others to prohibit human-nonhuman combinations through national legislation had met with no success up to 2009. Nevertheless, substantially the same bill has now passed into law at the state level in Louisiana. The Louisiana Senate Bill 115 follows the Draft Human-Animal Hybrid Prohibition Act of 2007 (S. 2358) in its definition of 'human-animal hybrid' except that it deletes from the first subcategory the phrase 'to render its membership in the species Homo sapiens uncertain'. It also omits the offence of transporting or receiving a human-animal hybrid and deletes the line 'or otherwise affecting interstate commerce'. (This last phrase had reflected a national rather than a state context).

The Bill was presented by Senator Daniel Martiny (R-Metairie), was signed by Governor Bobby Jindal in July 2009 and came into force in August 2009. The successful passage of this law at state level offers a model for other states. It is also an encouragement to those seeking to introduce such legislation at a national level. If the Louisiana law does become a model for national legislation then this would bring the United States into line with Argentina, Germany, Estonia, Switzerland, Canada, Australia and South Korea, though not with Japan.

In summary, from 1997 to 2009 a number of countries have brought in, or have attempted to bring in, legislation to place limits on the creation of human-nonhuman combinations. During this period one country has seemingly bucked the trend and brought in legislation expressly to facilitate the creation of such entities: the United Kingdom.

United Kingdom

The remainder of this chapter concerns the legislative situation in the United Kingdom. This is not because the United Kingdom offers an ideal model for international legislation. It does not. The legislation passed in the United Kingdom is at one end of a spectrum of possible approaches. Nevertheless, whether legislation is prohibitive (as Canada and Australia) or permissive (as Japan) it faces the same challenge of accurately defining terms. This is especially true for the UK model as research on human-nonhuman entities is only permitted under a licence. A key aim for legislators has therefore been to ensure that present and future research of this kind falls within the scope of licensing authorities. This requires the categories in UK law to be at least as clear and as comprehensive as they are in the legislation of Canada, or indeed of Louisiana in the USA.

The regulation of human-nonhuman combinations falls between legislation regulating research on nonhuman animals and legislation on human fertility and embryo research. But, as already mentioned, this may create some problems for embryos that could be considered under both human and nonhuman legislations or which are unregulated. As the UK Academy of Medical Sciences indicated in its 2011 report entitled 'Animals Containing Human Material':

> It is a matter of expert judgement to distinguish between embryos that are "predominantly human" and so come under the HFE Act [Human Fertilisation and Embryology Act 1990 which was revised in 2008], and embryos that are considered to be narrowly on the other side of the boundary and so "predominantly animal", and outwith the terms of the HFE Act.

The report added that in the UK 'These latter embryos are not currently regulated during early gestation (although their mothers are regulated under ASPA [Animals (Scientific Procedures) Act 1986])' (AMS 2011: 7–8).

In order to clarify this admission both sets of legislation will be examined in turn.

Animals (Scientific Procedures) Act 1986

The most important piece of legislation on research using nonhuman animals is the Animals (Scientific Procedures) Act 1986. As of 2012 this has not been substantially amended.

The Act is administered in England, Scotland and Wales by the Home Office and in Northern Ireland by the Department of Health, Social Services & Public Safety of the Northern Ireland Office. It provides information about the manner in which the Secretaries of State for the Home Department and for Northern Ireland propose to exercise their powers under the Act. Under the Scotland Act 1998, Schedule 5 (Reserved Matters), Part II (Specific Reservations), Head B (Home Affairs), B7 (Scientific procedures on live animals), the subject matter of the Animals (Scientific Procedures) Act 1986 is a reserved matter.

The Animal Procedures Committee advises the Home Office on animal experiments under this 1986 Act. The production of transgenic animals is covered by requirements of the Advisory Committee on Genetic Modifications (ACGM) and the Health and Safety Executive. They both require notification of the work. These controls derive from the Genetic Manipulation Regulations (1989) made under the power of the Health and Safety at Work Act 1974.

The Animals (Scientific Procedures) Act 1986 implements the requirements of the European Directive 86/609/EEC (outlined above under international legislation).

Section 1

(1) Subject to the provisions of this section, 'a protected animal' for the purposes of this Act means any living vertebrate other than man and any invertebrate of the species Octopus vulgaris from the stage of its development when it becomes capable of independent feeding.

(2) Any such vertebrate in its foetal, larval or embryonic form is a protected animal only from the stage of its development when:

(a) in the case of a mammal, bird or reptile, half the gestation or incubation period for the relevant species has elapsed; and
(b) in any other case, it becomes capable of independent feeding.

(3) The Secretary of State may by order:

(a) extend the definition of protected animal so as to include invertebrates of any description;
(b) alter the stage of development specified in subsection (2) above;
(c) make provision in lieu of subsection (2) above as respects any

animal which becomes a protected animal by virtue of an order under paragraph (a) above.

Section 2

(1) Subject to the provisions of this Section, 'a regulated procedure' for the purposes of this Act means any experimental or other scientific procedure applied to a protected animal which may have the effect of causing that animal pain, suffering, distress or lasting harm.[3]

(2) An experimental or other scientific procedure applied to an animal is also a regulated procedure if:

(a) it is part of a series or combination of such procedures (whether the same or different) applied to the same animal; and
(b) the series or combination may have the effect mentioned in subsection (1) above; and
(c) the animal is a protected animal throughout the series or combination or in the course of it attains the stage of its development when it becomes such an animal.[4]

(3) Anything done for the purpose of, or liable to result in, the birth or hatching of a protected animal is also a regulated procedure if it may as respects that animal have the effect mentioned in subsection (1) above.

---

[3] Anything done for the purpose of, or liable to result in, the birth or hatching of a protected animal that may as a result of the procedure experience pain, suffering, distress or lasting harm is also a regulated procedure [Section 2(3)]. Thus,
– the breeding of animals with harmful genetic defects;
– the manipulation of germ cells or embryos to alter the genetic constitution of the resulting animal; and
– the subsequent breeding of such genetically modified animals
are regulated if the intention is to maintain the animals produced beyond midway through gestation or incubation.
In Guidance on the Operation of the Animals (Scientific Procedures) Act 1986, May 2000, paragraph 2.18.
[4] Under Section 2(2)(c), protection is also provided when regulated procedures are applied at an earlier stage of development if:
(i) the animal is to be allowed to live beyond the stage of development set out in paragraph 2.7 above; and
(ii) the procedure may result in pain, suffering, distress or lasting harm after the animal has reached that stage of development.
For example, licences are required for virus propagation in embryonated bird eggs if inoculation takes place before the mid-point of incubation and if the embryo is allowed to survive into the second half of the incubation period.

Section 3

No person shall apply a regulated procedure to an animal unless:

(a) he holds a personal licence qualifying him to apply a regulated procedure of that description to an animal of that description;

(b) the procedure is applied as part of a programme of work specified in a project licence authorising the application, as part of that programme, of a regulated procedure of that description to an animal of that description; and

(c) the place where the procedure is carried out is a place specified in the personal licence and the project licence.

Section 4

(1) A personal licence is a licence granted by the Secretary of State qualifying the holder to apply specified regulated procedures to animals of specified descriptions at a specified place or specified places.

Section 5

(1) A project licence is a licence granted by the Secretary of State specifying a programme of work and authorising the application, as part of that programme, of specified regulated procedures to animals of specified descriptions at a specified place or specified places.

The definition of a 'protected animal' according to this legislation to include 'any living vertebrate other than man' is directly parallel to the 1986 European Directive and suffers from the same defect. How is it to be decided if an entity is 'other than man'? How human does an entity have to be before it falls outside this legislation and within the scope of law specifically concerned with human welfare? In practice the scope of the Animals (Scientific Procedures) Act 1986 is resolved not by its own provisions but by the scope of the relevant human-focused legislation. The provisions of the Animal Act thus cover those entities which are not human according to the law that concerns research on human beings.

In relation to embryo research (which is currently the focus of research interest in human-nonhuman combinations), the Animals (Scientific Procedures) Act 1986 may then address some residual categories for what does not fall under the Human Fertilisation and Embryology Acts of 1990 and 2008. It is therefore necessary first to consider these Acts and then return to the implications of this for the application of the Animals (Scientific Procedures) Act 1986.

Human Fertilisation and Embryology Act 1990

Section 1. (1) In this Act, except where otherwise stated –

(a) embryo means a live human embryo where fertilisation is complete, and

(b) references to an embryo include an egg in the process of fertilisation,

and, for this purpose, fertilisation is not complete until the appearance of a two cell zygote.

Section 3. (2) No person shall place in a woman –

(a) a live embryo other than a human embryo, or

(b) any live gametes other than human gametes.

Section 3(3)(b) A licence cannot authorise placing a [human] embryo in any animal.

Section 4. (1) No person shall:

mix gametes with the live gametes of any animal, except in pursuance of a licence.

Schedule 2, Section 1. (1)(f) also indicates that a licence may authorise, in the course of providing treatment services:

mixing sperm with the egg of a hamster, or other animal specified in directions, for the purpose of testing the fertility or normality of the sperm, but only where anything which forms is destroyed when the test is complete and, in any event, not later than the two cell stage.

Under the Scotland Act 1998, Schedule 5 (Reserved Matters), Part II (Specific Reservations), Head J (Health and Medicines), J3 (Embryology, surrogacy and genetics), the following are reserved matters to the UK parliament:

- The subject-matter of the Human Fertilisation and Embryology Act 1990, and

- Human genetics.

Before the 1990 Act was prepared there had been little national or intentional discussion on research involving human-nonhuman combinations and the law's treatment of them was cursory. The only procedure which was directly addressed under this legislation was the mixing of human gametes with the live gametes of 'a hamster, or other animal specified in directions' as a test for sperm fertility (Section 4. (1) (c)). A licence could be obtained for this procedure but the product of this technique had to be destroyed at or before the two cell stage. It is this provision which would shape the Australian law of 2006.

In the wake of scientific developments with SCNT and legislative moves in Canada and elsewhere it became obvious that there was an urgent need for new UK legislation. Under the 1990 Act the creation of many human-nonhuman entities would have existed in a legal vacuum. This was confirmed by the House of Commons Science and Technology Committee. Its 2005 report Human Reproductive Technologies and the Law indicated that the consideration of human-nonhuman chimeras and hybrids was made difficult by the lack of legal definitions (HCSTC 2005: 30–2).

For example, since the Human Fertilisation and Embryology Act 1990 mentions the mixing of human and animal gametes, it may be assumed to have been drafted with the possibility of addressing human-nonhuman combinations. However the 1990 Act did not provide adequate clarifications concerning the status of these entities. Because of this situation, it was also uncertain whether the UK Human Fertilisation and Embryological Authority (HFEA) was even entitled to regulate the creation of most of the human-nonhuman entities being envisaged in research.

In other words, whether or not these entities came under the jurisdiction of the HFEA depended on whether or not they were 'human', but this was not specified in the legislation. It is probable that controls relating to the creation of such human-nonhuman entities would only exist if the HFEA was entitled to adopt a broad definition of a 'human' embryo. However, this would then be open to a legal challenge and the decision might then have to be resolved in the courts. This was the route by which research on cloned embryos was incorporated into the legislation after the ruling of R (Quintavalle) v. Secretary of State for Health of the 13 March 2003. In this House of Lords (HL) decision, Lord Bingham of Cornhill indicated that Parliament could not have intended to distinguish between embryos produced by, or without, fertilisation since it was unaware of the latter possibility. The reference to fertilisation was not therefore integral to the definition but was directed to the time at which an embryo should be treated as such. However, no such interpretation is automatic and the HFEA should address this issue as soon as possible in order to clarify the matter.

Human Fertilisation and Embryology Act 2008

The Department of Health reviewed the Human Fertilisation and Embryology Act 1990 and in December 2006 came up with recommendations for revised legislation. In relation to 'Embryos containing human and non-human material' the review concluded that the government would 'propose that the creation of hybrid and chimera embryos *in vitro* should not be allowed'. However, it also proposed that 'the law will contain a power enabling regulations to set out circumstances in which the creation of hybrid and chimera embryos *in vitro* may in future be allowed under licence'. This was ostensibly a middle position. It purported to prohibit human-nonhuman combinations but in fact allowed regulations to permit the creation of these entities without having to pass new primary legislation. Thus, in comparison with the Australian and Canadian law or the draft legislation in the United States this proposal was decidedly permissive.

The British government was therefore taken by surprise by an extraordinarily fierce backlash from scientific lobby groups which orchestrated a sustained media campaign on the issue. A campaign that was organized primarily through the services of the Science Media Centre which exists to promote the views of the said scientific community to the national news media. Typical of the early coverage was an article by Mark Henderson, Science Editor of the Times, warning that, 'Patients with incurable crippling diseases may be denied the first effective treatments because of government plans to outlaw the creation of 'human-animal' embryos' (Henderson 2007). This extraordinary claim of 'effective treatments' for current patients cannot be disguised by the use of the word 'may'. Indeed in future reporting there was a tendency for this 'may' to become a 'will'. For example, in articles by the same author it was mentioned that 'Medical researchers ... think this will help them to understand diseases such as Parkinson's and Alzheimer's' (Henderson 2008a) or 'British scientists will be allowed to research devastating diseases such as Alzheimer's and Parkinson's using human-animal embryos' (Henderson 2008b).

The campaign to permit human-nonhuman combinations used patient groups who had been persuaded that human-nonhuman combinations were essential to medical progress for their disease (Cobbe 2011). The success of this tactic in 2008 was seen in particular with the Prime Minister, at the time, Gordon Brown for whom the issue may have resonated in a deeply personal manner. His son had cystic fibrosis, and this was one of the conditions which, it was alleged, 'could one day benefit from embryo research' and that 'only hybrids can help solve problems' that currently inhibit this research (Hinsliff 2008). It was perhaps this emotional investment in the legislation, in addition to other more pragmatic reasons, which may explain why Gordon Brown at first refused to follow the precedent of the 1990 Act and allow a 'conscience vote' on those parts of the Act that covered embryo research. It was only after a letter to the Times signed by over one hundred professors, and after the threat of rebellion by members of the cabinet that he relented and allowed a free vote. Even the leader of the British Conservative Party at the time, David Cameron, who then became Prime Minister, was led to believe in May 2008, that his severely disabled son may benefit from the results of human-nonhuman embryonic research. This was particularly tragic since his son then died less than a year afterwards (Winnett 2008).

Throughout 2007 and 2008 the United Kingdom witnessed an intense media campaign on this issue with a lot of unreasonable exaggeration concerning the benefits of human-nonhuman research being presented in an effort to overcome initial public qualms (Cobbe 2011). However, while this strategy was rhetorically very successful, it did a disservice to the public understanding of science. For example, of the several dozen briefings and press releases from the Science Media Centre which coordinated the scientific lobby, not one alerted the public to doubts among some scientists that human-nonhuman combinations were as promising as was being claimed. A responsible balance in reporting was non-existent and this was deliberate. Reminiscing on the lead-up to the UK's human-nonhuman interspecies legislation, Mark Henderson of *The Times* indicated in 2009 that,

> it's my job to analyse and interpret science in the news, and in this case there was no scientific reason for a ban. If reflecting that means I'm partisan, then I cheerfully plead guilty. While fairness is essential to good scientific reporting, an overemphasis on balance can be its enemy. (Henderson 2009)

There was thus widespread public astonishment when immediately after the successful passage of the new law, and even before it came into force, the only two grant applications for research in this area were turned down by the UK Medical Research Council which provides government grants in biomedicine. The supposedly 'vital' avenue of research had already been eclipsed by better alternatives (Sample 2009).

The public and parliamentarians were thus very poorly served by the media on the content of the 2008 Act. There was very little reflective discussion of the perplexing nature of human-nonhuman combinations. The debate was rather conducted as a battle between miracle cures and minotaurs, neither of which represented reality. Nevertheless, a crude level of political debate in the popular media is quite compatible with careful provisions within an act itself. While parliamentarians disagreed about which categories of human-nonhuman combination should be permitted, all were agreed that the law should define categories that were as clear and as comprehensive as possible. In relation to the question of definitions the two most important sections of the Act are Section 1 and Section 4. It is to these definitions that we now turn.

Section 1: Meaning of 'embryo' and 'gamete'.

1) Section 1 of the 1990 Act (meaning of 'embryo', 'gamete' and associated expressions) is amended as follows.
2) For subsection (1) substitute—

(1) In this Act (except in section 4A or in the term 'human admixed embryo')—

(a) embryo means a live human embryo and does not include a human admixed embryo (as defined by section 4A(6)), and
(b) references to an embryo include an egg that is in the process of fertilisation or is undergoing any other process capable of resulting in an embryo.

An embryo is defined under the new section 1 (1) in broad terms as a 'live human embryo' but the definition no longer assumes that an embryo can only be created by fertilization. This clarifies that cloned human embryos, produced by SCNT are 'embryos' covered by the 2008 Act.

The definition of an 'embryo' in the new section 1 (1)(a) of the 1990 Act excludes certain types of embryos created by combining human and nonhuman material. In an earlier draft of this legislation these entities were referred to as 'interspecies embryos' but in the 2008 Act, as passed, they are referred to as 'human admixed embryos'. These entities are not prohibited, but they constitute a different category from purely human embryos. For example, human admixed embryos can legally be produced for research purposes but they cannot be implanted into a woman or into an animal. They are considered in detail under Section 4A below.

Later in Section 1, the 1990 Act is amended to expressly encompass not only mature eggs and sperm, but also immature gametogenic cells such as primary oocytes, and spermatocytes. There is also a regulation-making power enabling the Secretary of State to expand the definitions of 'embryo', 'eggs', 'sperm' or 'gametes', where this is considered to be necessary or desirable in light of developments in science or medicine.

The meaning of permitted eggs and permitted embryos is extended to include eggs or embryos that have been treated in such a way as specified in regulations to prevent the transmission of serious mitochondrial disease. Mitochondria are found outside the nucleus of the cell and contain a small amount of DNA. They are involved in energy production and are present in most cells in the body. If a woman's egg is fertilized by sperm the mitochondria from her egg will become the mitochondria for every cell of the embryo formed. Therefore, if a woman has a genetic medical condition associated with her mitochondria, these will be inherited via her eggs. Some mitochondrial diseases are also caused by defects in nuclear DNA (since some mitochondrial proteins are encoded for by nuclear genes). In these cases, the mitochondrial disease could be inherited from either mother or father.

This is an example of the way the government has sought to 'future-proof' the legislation. This kind of provision helpfully allows a future government to react quickly to new developments not envisaged in the legislation. On the other hand, by devolving the power to the Secretary of State this reduces the scrutiny and democratic accountability of the law.

It is in Section 4 that the Act turns to the topic of human-nonhuman combinations:

Section 4: Prohibitions in connection with genetic material not of human origin

(2) After section 4 of the 1990 Act insert—

4A Prohibitions in connection with genetic material not of human origin
(1) No person shall place in a woman—

(a) a human admixed embryo,
(b) any other embryo that is not a human embryo, or
(c) any gametes other than human gametes.

(2) No person shall—

(a) mix human gametes with animal gametes,
(b) bring about the creation of a human admixed embryo, or
(c) keep or use a human admixed embryo,
except in pursuance of a licence.

(3) A licence cannot authorize keeping or using a human admixed embryo after the earliest of the following—

(a) the appearance of the primitive streak, or

(b) the end of the period of 14 days beginning with the day on which the process of creating the human admixed embryo began, but not counting any time during which the human admixed embryo is stored.

(4) A licence cannot authorise placing a human admixed embryo in an animal.

(5) A licence cannot authorise keeping or using a human admixed embryo in any circumstances in which regulations prohibit its keeping or use.

(6) For the purposes of this Act a human admixed embryo is—

(a) an embryo created by replacing the nucleus of an animal egg or of an animal cell, or two animal pronuclei, with—

(i) two human pronuclei,
(ii) one nucleus of a human gamete or of any other human cell, or
(iii) one human gamete or other human cell,

(b) any other embryo created by using—

(i) human gametes and animal gametes, or
(ii) one human pronucleus and one animal pronucleus,

(c) a human embryo that has been altered by the introduction of any sequence of nuclear or mitochondrial DNA of an animal into one or more cells of the embryo,

(d) a human embryo that has been altered by the introduction of one or more animal cells, or

(e) any embryo not falling within paragraphs (a) to (d) which contains both nuclear or mitochondrial DNA of a human and nuclear or mitochondrial DNA of an animal ('animal DNA') but in which the animal DNA is not predominant.

(7) In subsection (6)—

(a) references to animal cells are to cells of an animal or of an animal embryo, and

(b) references to human cells are to cells of a human or of a human embryo.

(8) For the purposes of this section an 'animal' is an animal other than man.

The law prohibits some actions absolutely, such as the placing of a human admixed embryo into a woman (4A (1)). It prohibits other actions only conditionally, where no licence has been obtained. This is the case with the mixing of human and nonhuman gametes, and indeed with the creation of all categories of human admixed embryos (4A (2)). It might be argued that the requirement for a licence creates a robust safeguard which could reassure the public that inappropriate experimentations would not be carried out. However, during the passage of the bill, the Secretary of State for Health admitted that in almost 20 years of operation, the Human Fertilisation and Embryology Authority had never in fact refused a research licence (all but one being granted in the first instance and one being granted on appeal). In practice prohibitions 'except in pursuance of a licence' may not offer effective or appropriate safeguards. These are only created by the bright lines of exceptionless prohibitions.

The key paragraph which defines human admixed embryos is 4A(6). This identifies four kinds of human-nonhuman combination plus a fifth catch-all category.

In terms of the definitions offered in the present book, 4A (6)(a) refers to three kinds of cybrids (in each case all the nonhuman nucleus is removed and 100 per cent of the nuclear DNA in the resulting entity will be human). 4A (6)(b) refers to a true hybrid. 4A (6)(c) refers to a transgenic individual. 4A (6)(d) refers to a chimera. 4A (6)(e) is a catch-all which makes clear that the overall category is an embryo that contains both human and nonhuman DNA and where the human DNA predominates.

In specifying that the human DNA must be predominant the legislation excludes all human-nonhuman combinations which are predominantly nonhuman. Hence, for example, 4A (6)(c) and 4A (6)(d) describe modifications to a human embryo by adding nonhuman DNA or nonhuman cells.
However neither section describes a nonhuman embryo that has been modified through the addition of human DNA or cells. These entities would not be human admixed embryos and would not fall under the Act.

It is significant that the first kind of human-nonhuman combination described is a cybrid, and that three different kinds of cybrids are described. The debates that shaped the law were largely concerned with the alleged promise of research on cybrids. In this case the nuclear DNA is all from a human source and it is natural to regard these embryos as modified human embryos. As a point of principle they do represent a mixing of human and nonhuman material. They contain nonhuman DNA in the cytoplasm (inside tiny organelles called mitochondria). Nevertheless they are predominantly human and if a cybrid child could be produced and survive to term then he or she undoubtedly would be considered to be a human being for legal purposes. It is this example legislators had primarily in mind when coining the phrase 'human admixed embryo'. This explains why the first adjective is 'human' and why the term 'interspecies' embryo was felt to be inappropriate.

In contrast, very little time during the public debate was spent considering the more perplexing cases of true hybrids or chimeras. In these cases it is not clear that the human characteristics would predominate. They rather represent examples which could be so mixed that one is perplexed as to what status they have. If a humanzee were to be born it is not clear whether he or she would be protected as a human. The term 'human admixed' does not seem very helpful for this more evenly mixed kind of human-nonhuman combination. The law was drafted with more attention to cybrids, which are at the ends of the spectrum (mostly human or mostly nonhuman) than to true hybrids which are in the middle (roughly half human, half nonhuman), or chimeras which could be anywhere on the spectrum. The resulting definitions thus leave some of the most problematic combinations outside the legislative framework. For example, it is not clear why creatures which are evenly mixed but where the nonhuman element is slightly 'predominant' should be excluded from the legislation.

It is regrettable that the legislation uses animal to mean a nonhuman animal (4A (8)). This was avoided in the Canadian legislation which could have been used as a model. Nevertheless, at least the term 'human-animal hybrid' was not chosen as the overarching category (as it was, for example, in the draft American legislation). The importance of recognizing that human beings are animals of a specific kind will be discussed further in the third part of this book in the context of the cultural, ethical and world view perspectives on human-nonhuman combinations.

In considering the concept of the predominance of human or nonhuman DNA it can also be questioned whether this can ever have any exact meaning since, as Robert and Baylis indicate 'Much of 'our' DNA is shared with a huge variety of apparently distantly related creatures (e.g. yeast, worms, mice, etc.)' (Robert and Baylis 2003). It is, therefore, difficult to specify how much DNA is *uniquely* human.

Later in section 4 (subsection 9 and 10) the 1990 Act is again amended to expressly encompass embryos created other than by fertilization and to encompass not only mature eggs and sperm, but also immature gametogenic cells such as primary oocytes, and spermatocytes. There is also (in subsections 11 and 12) another regulation-making power similar to that in section 1. This also is an attempt to 'future-proof' the legislation.

Further explanations relating to the creation on human-nonhuman embryonic combinations in the HFE Act 2008

In the final House of Lords discussions (29 October 2008), the Government Minister, Lord Darzi of Denham, gave some further explanations concerning the creation of human-nonhuman embryonic combinations (*Lords Hansard*, 29 October 2008, Column 1624). Since these explanations from a government minister were given just before the final vote on the legislation, they have the force of an official explanation with respect to the intention of the provisions should they eventually be contested in the UK courts.

Human admixed embryos include embryos in which the human DNA may predominate at any stage during the entire period of the embryo's existence (even if the human DNA does not predominate at the beginning)

In this regard, Lord Darzi explained that:

> If it were considered that an embryo was to be created in which the human DNA would ultimately predominate, an application for an admixed research licence would have to be made to the HFEA at the outset. This is because a licence is required to bring about the creation of a human admixed embryo. If a researcher was intending to create an embryo that would at some stage be predominantly human, for however short that time might be, they would need a licence to do so.

> The noble Lord, Lord Walton, also referred to tetraploid complementation where the cells of an early animal embryo are altered so that they contain twice the usual complement of DNA. These cells could give rise only to extra-embryonic tissue—for example, a placenta—and any human cells placed within it could give rise to the embryo proper. It would be an admixed embryo for the purposes of the new catch-all category that the Bill is adding to the definition of human admixed embryo at Clause 4. (*Lords Hansard*, 29 October 2008, Column 1624)

This means that the word 'predominant' covers the whole period of an embryo's existence. In other words, any chimeric embryos in which the animal cells first predominate but where the human cells eventually develop into a majority – for however brief a time – would be covered by the catch-all category of the 2008 Act (Section 4A (6)(e)) and would need a licence from the HFEA. These include tetraploid complementation embryos. These are embryos in which a few human pluripotent stem cells are placed inside an animal embryo in which all the cells are tetraploid with four sets of chomosomes instead of just two.[5] In research, it is often impossible to know in advance whether a chimeric embryo in which the animal cells first predominate might eventually become an embryo in which the human cells predominate. In other words, if an animal-human chimeric embryo begins its development with a majority of animal cells, it will be unregulated under human legislation and could be left to develop well past the 14-day limit.

Human admixed embryos include those embryos in which human 'functionality' predominate

The Government Minister, Lord Darzi of Denham, also indicated on the same day that:

> In the case of an embryo in which the brain might be predominantly animal, it is worth reminding ourselves what we mean by 'predominant'. We refer not only to the percentage of the DNA but also to its location and functionality. If that entity had a human brain, that could clearly have a predominant function so, by definition, it would be at the human end of the spectrum of human admixed embryos and would require an HFEA licence'. (*Lords Hansard*, 29 October 2008, Column 1625)

In this respect, and in a reply to Lord Elton who asked whether predominance would not only depend upon percentage but also on location or function and whether that meant that if the cells concerned relate to an organ, such as the brain, although they are less than 50 per cent, they still predominate, Lord Darzi of Denham replied:

> [T]he answer is yes. If that functionality predominates being human, that embryo will be classified as being at the human end of the spectrum of human admixed embryos'. (*Lords Hansard*, 29 October 2008, Column 1626)

But this then raises the question about the manner in which a 'functionality' is defined, to which the UK Minister of Health responded (in a letter from the Minister, 22 May 2009), by indicating that:

---

[5] By applying a direct current pulse, a single cell may be created through the fusion of two blasto-meres (i.e. early cells) at the two-cell stage of an embryo. The replication of the genetic material followed by mitotic division results in a two-cell embryo containing double the diploid (two sets of chromosomes) content of DNA. This tetraploid embryo can develop further to the blastocyst stage.

The point that Lord Darzi was making is that it is not necessarily the amount of DNA that would determine predominance, but what the DNA actually does ... Each case would have to be considered separately. It will be for the Human Fertilisation and Embryology Authority to apply the test in practice, and ultimately for the courts to determine in law what this test should be in any particular case.

This means that if all the different parts of an embryo had a majority of animal DNA but that some parts, such as the brain, still started to behave as predominantly human (even though the majority of the DNA in the brain was animal), then this embryo could be considered as a human admixed embryo and could come under the remit of the HFE Act 2008. However, this determination is not on the face of the legislation and so would be reliant on a decision by the Human Fertilisation and Embryology Authority (or any other new body with the same remit), a decision which might well be open to legal challenge.

It was commendable that the government intended to include in its legislation entities which were majority nonhuman by sheer mass but where a human functionality is predominant, as for example with the brain. However it remains unclear what is meant by 'human functionality'. How does one know whether or not an embryo has a predominant human functionality? Again it also raises the question of why the test should be that, even on the level of function, the entity should be predominantly human. For example, if an animal has a canine brain but has some other identifiable human functionality (for example the ability to produce human gametes) is this not disturbing? Is it necessary to say that the animal is 'predominantly human' before it falls under appropriate legislation?

This was a quandary that was also raised by the Academy of Medical Sciences' 2011 report entitled 'Animals Containing Human Material'. It indicated (AMS 2011: 89) that under the current UK legislative framework, in such uncertain situations where extensive mixtures of nonhuman and human DNA are present, it would be necessary for scientists to either:

Hold licences for the research under both human and nonhuman legislation from the outset of the experiment.

or,

Ensure, through close monitoring, that the experiment was immediately halted if the outcome was unexpected and the experiment was being conducted solely under a nonhuman legislation and it became evident that the human threshold had been reached. Further authorization from the body regulating human legislation should then be sought before resuming the research.

This solution, however, may be considered as unacceptable and impractical from an ethical perspective. It is unfortunate that, despite repeated warnings concerning this human-nonhuman boundary dilemma during the preparation of legislation, the UK parliament did not see fit to appropriately address the problem (HCSTC 2007, HC272-II, Memorandum 9, Submission form the Scottish Council on Human Bioethics, Evidence 56–61). As the Academy of Medical Sciences emphasized, 'the UK legislative structure is such that some awkward cases may fall at the boundary of jurisdiction' (AMS 2011: 113). It also noted that there is a real need for a process to exist which is as smooth and clear as possible which ensures 'that there are no regulatory gaps, overlaps, or inconsistencies', between human and nonhuman legislations and avoids any chance that contentious experiments might evade appropriate scrutiny (AMS 2011: 113).

Animals (Scientific Procedures) Act 1986

With these concerns in mind it is useful to return to consideration of the Animals (Scientific Procedures) Act 1986 as it is this Act that must cover entities that fall outside the Human Fertilisation and Embryology Acts of 1990 and 2008. The 1986 Animals Act gives some powers to the Secretary of State to extend the definition of a protected animal (Section 1(3)(c)). Nevertheless, in a personal communication to Dr Calum MacKellar from the UK Home Office on 2 April 2007 it is indicated that 'An embryonic animal-human chimera that is destroyed before the gestation or incubation period has elapsed does not come under the remit of the 1986 Act. Consequently there are no orders from the Secretary of State'.

Moreover, in a written question, on 7 January 2008, Lord Alton of Liverpool asked Her Majesty's Government:

Further to the Written Answer by Lord Darzi of Denham on 18 December 2007 (WA 116), whether they will specify the relevant provisions of legislation that would prevent the Home Office from permitting the implantation into an animal of any interspecies embryo in which the animal cells primarily produce extra-embryonic tissues.[6] (HL1150)

---

[6] Written questions in Parliament are given a number, by convention cited in square brackets at the end of the question. For questions in the House of Lords they begin HL, hence [HL1150].

In response, Lord Darzi of Denham indicated (*Lords Hansard*, 30 Jan 2008: Column WA121-2) that:

> Before granting a license the Secretary of State is required to consult one of the Home Office inspectors appointed under the Animals (Scientific Procedures) Act 1986 (ASPA). The Secretary of State may also seek advice on applications from an independent assessor and/or the Animal Procedures Committee (the statutory advisory committee established by ASPA).
>
> A number of categories of licence application are automatically referred to the Animal Procedures Committee for advice. These include applications of any kind raising contentious issues, or giving rise to serious societal concerns (for example involving the creation of embryos through tetraploid complementation using human material).
>
> Advice is sought from independent assessors in cases involving highly specialised science, and where there are significant animal welfare concerns or concern as to the likelihood of success or about the approach being taken.
>
> Section 5 of ASPA allows the Secretary of State discretion when reaching licensing decisions. It is not sufficient that an application satisfies the requirements of the Act for it to be granted. Using this discretion, the Secretary of State can adopt policies that certain categories of work will not be licensed. There are several examples of this such as the use of great apes and the testing of alcohol and tobacco products.

In this regard, it should be noted that it is the UK Home Office policy and practice to refer all cases that may be sufficiently novel or could potentially raise societal concerns to the Animal Procedures Committee for advice. This is then sent to the Secretary of State at the Home Office who may also receive advice from the Animals Scientific Procedures Inspectorate and independent experts.

Interestingly, in a 2001 report, the Animal Procedures Committee recommended that 'No licences should be issued for the production of embryo aggregation chimeras especially not cross-species chimeras between humans and other animals, nor of hybrids which involve a significant degree of hybridization between animals of very dissimilar kinds' (Animal Procedures Committee 2001, Recommendation 5, Paragraph 57). This recommendation, however, is not binding and it remains to be seen whether the proposals remain valid.

Moreover, it should be emphasized that the creation of an embryo containing human and nonhuman DNA, where the nonhuman DNA predominates, would be referred to the Animal Procedures Committee if any regulated procedure involving a live protected animal, for example, superovulation to obtain eggs, is involved. However, where the source material, such as animal eggs and/or DNA, could be obtained without performing any regulated procedure to a living protected animal (for example the use of post-mortem material removed from humanely killed animals), such work could be undertaken (in cases where the embryo would neither develop to halfway through the gestation periods of the animal in question, nor be implanted/transferred into any protected living animal) without any authorities being required under the 1986 Act.

In such cases, there would be no request made to the UK Home Office for authorities under the 1986 Act, and therefore no proposal to forward to the Animal Procedures Committee for advice (personal communication to Dr C MacKellar from the UK Home Office, 20 April 2009). This would happen even if the created embryo contained a mixture of human and nonhuman primate DNA in which the nonhuman primate DNA predominated.

In addition, it should be noted that the placement of any kind of embryo into the uterus of a protected animal is automatically a regulated procedure for the purpose of the Animals (Scientific Procedures) Act 1986 if it is undertaken for an experimental or other scientific purpose whether or not the pregnancy is subsequently terminated (House of Lords, *Hansard,* Written Answers, 25 Jun 2007, Column WA103). This is the case regardless of whether the implantation required a surgical procedure – not only because of the nature of the intervention required to implant the material, but also because of the possible welfare problems that might arise post-implantation (personal communication from the UK Home Office, 20 April 2009).

Moreover, any chimeric embryo which does not come under human legislation and is placed into the uterus of an animal that develops beyond halfway through gestation will also itself be deemed to be a protected animal having undergone regulated procedures (House of Lords, *Hansard,* Written Answers, 25 Jun 2007, Column WA103). This means that before any implantation for gestation of an animal-human embryonic combination into an animal takes place, expert professionals would need to be consulted. However, once the due process has been undertaken, a licence could still be given for such a procedure.

In short, the provisions of the 1986 Act give the Secretary of State power to oversee research using human-nonhuman combinations. Nevertheless, these provisions would not cover research until and unless the embryo was implanted. Furthermore, even where the Secretary of State and the Animal Procedures Committee were involved, they would not benefit from guidance or limits explicit in the legislation. In other words, while the 1990 Act has been revised the 1986 Act is in need of revision so that it deals explicitly with human-nonhuman combinations.

The 2011 Academy of Medical Sciences' report on animals containing human material

During the discussions that were taking place in the preparation of the revised Human Fertilisation and Embryology Act 2008, which looked at human-nonhuman embryos at the human end of the spectrum, it became quickly evident that new guidelines were also necessary in the UK addressing the converse situation i.e. interspecies embryos and other entities which were at the nonhuman animal end of the spectrum. These had, indeed, received relatively little consideration even though they could present regulatory and ethical challenges. As a result, the Academy of Medical Sciences began preparing a report in 2009 which was eventually published under the title 'Animals Containing Human Material' in July 2011. Of course, some of the entities being created falling in this category are not new and thousands have already been used in biomedical research but they have received relatively little public discussion. Their role, moreover, is becoming increasingly important in the study of diseases and the development of new treatments.

The report was prepared to review the UK regulatory environment in this rapidly developing area of research on nonhuman embryos and animals containing human material while examining the scientific, social, ethical and safety aspects. The recommendations could then be used to begin and develop an engagement with the general public on the relevant issues.

At the end of the Academy's report (which has no legal weight) the Academy of Medical Sciences proposed a number of categories in which experiments involving animals containing human material could be classified (AMS 2011: 110–11). These were:

Category 1
The great majority of experiments in which animals containing human material do not present issues beyond those of the general use of animals in research. These should be subject to the same oversight and regulation under Animals (Scientific Procedures) Act 1986 as other animal research.

Category 2
The limited number of experiments with animals containing human material which should be permissible subject to additional specialist scrutiny on a case-by-case basis by an expert body. This should happen at least until experience allows the formulation of guidelines. Strong scientific justification should be

provided to this expert body, which should closely consider the ethical and safety issues in addition to the potential value of the research. Although the Academy would expect experience to evolve over time, the major types of research that would currently be included under this category would be:

Substantial modification of an animal's brain that may make the brain function potentially more like a human brain, particularly in large animals.

Experiments that may lead to the generation or propagation of functional human germ cells in animals.

Experiments that could be expected to significantly alter the appearance or behaviour of animals, affecting those characteristics that are perceived to contribute most to distinguishing the human species to other species that are closely similar.

Experiments involving the addition of human genes or cells to nonhuman primates.

Category 3

The very small number of experiments which should not be possible, at present, because they lack compelling scientific justification or raise very strong ethical concerns. The Academy suggests that a list of such experiments should be kept under regular review by an expert body, but should include:

Allowing the development of an embryo, formed by the combination of nonhuman primate and human embryonic or pluripotent stem cells before implantation, beyond 14 days of development or the first signs of primitive streak development, (whichever occurs first), unless there is convincing evidence that the fate of the implanted (human) cells will not lead to 'sensitive' changes in the observable characteristics or traits of the developing foetus.[7]

---

[7] This supplements the 14-day provision applied to human admixed embryos under the Human Fertilisation and Embryology Act 2008, so that mixed embryos that are considered to not quite meet the criteria for being 'predominantly human', should nevertheless be regulated on the basis of the likely characteristics and traits of the embryos created. At present, any mixed origin embryo considered to be 'predominantly human' is regulated by Human Fertilisation and Embryology Act 2008 and cannot be kept beyond the 14-day stage, whereas an embryo considered to be predominantly animal is unregulated until the mid-point of gestation. As to whether or not an admixed embryo is predominantly 'human' is an expert judgement and will not always be easily predictable in the current state of science.

Transplantation of an amount of human-derived neural cells into a nonhuman primate sufficient to make it possible, in the judgement of an expert body, for the primate's brain to be so substantially modified in its function as to engender 'human-like' behaviour.

Breeding of animals that have, or may develop, human-derived germ cells where this could lead to the production of human embryos or true hybrid embryos within an animal.

Conclusion

In summary it can be noted that even in the United Kingdom, which has had a national debate and enacted legislation in 2008 explicitly to deal with human-nonhuman combinations, there remain serious gaps in the law. As indicated by Haddow et al., 'there was, and continues to be, an apparent lack of reflective, explicit, and integrated regulatory approaches to admixed entities, e.g. a certain level of ad hoc-ery is disclosed, that is catalyzed by scientific innovation' (Haddow et al. 2010). This situation in part can be traced back to the poor quality of the public debate, in inverse proportion to its emotional intensity. The media coverage was focused only on one kind of combination and its allegedly promise in medical research. What was lacking was a broader and more reflective discussion about different kinds of human-nonhuman combination.

The situation of the UK, however, is not unique. It is clear that very little discussion has taken place in a large majority of countries, which do not have any adequate legislation in the field of interspecies entities. This may be because of a lack of foresight or even because of ignorance on the part of national policy-makers (Taupitz and Weschka 2009: 451).

In the light of these considerations there is evidently a need, for example, for the National Ethics Committees of the Council of Europe member states to initiate an extensive consultation and reflection on the complex ethical questions arising from the creation of human-nonhuman combinations. This could then help shape national legislation and pave the way for the negotiation of international conventions, whether at a European level or through the United Nations. If this is not undertaken then most states will be unprepared to cope with new developments in this area.

Patenting animal human combinations

Patent legislation has long contained provisions allowing patents to be refused on public policy grounds. For instance, in the United Kingdom, Section 1(3) of the Patents Act 1977 as originally enacted provided that a patent should not be granted for an invention which is 'likely to encourage offensive, immoral or antisocial behaviour'; similarly, in the United States, there has traditionally been a similar exclusion for inventions which are 'frivolous or injurious to the well-being, good policy, or sound morals of society' (Lowell v. Lewis, 15 (a. 1018 No. 8568) (C.D. mass. 1817)).

The origin of these exclusions is the common law concept of public policy, by which is meant not government policy, but, rather what is offensive to the sense of morality in a society.

Napoleonic systems have a close analogue of the common law concept of public policy in the notion of 'ordre public' which, however, is rooted in a philosophy which is not strictly the same: putting emphasis on the good order of society rather than morality as such. However, concepts of public policy and of ordre public are constantly changing in tune with the mores of society.

Another concept which is determinative in the patenting process is that of 'humanity' since many countries, such as Canada, the USA and Europe prohibit the patenting of human beings.

However, a lot of uncertainty remains relating to exactly what is defined as a human being. Moreover, other problems exist since there are sometimes no qualitative features, such as morphological, genetic, or behavioural features, that can be considered essential to species membership. In addition, there is still no societal consensus relating to the beginning of individual human or nonhuman animal life (Ereshefsky 2007).

Finally, there is also uncertainty relating to the biological concepts of 'higher' and 'lower' life forms, which reflect their degrees of cellular complexity. This is very important for some patent regulations, such as in Canada, where all higher life forms, including humans, are unpatentable (Hagen and Gittens 2008: 23; Harvard College, [2002] 4 S.C.R. 45, 2002 SCC 76 ¶155 (Can.)).

This has led to some commentators, such as Gregory Hagens and Sébastien Gittens, indicating that 'any distinction to be drawn between human and non-human based upon relative amounts of human and non-human DNA would appear to be arbitrary' (Hagen and Gittens 2008: 21). They go on to indicate that:

> In terms of mixed species, unlike the use of humanity as a criterion, it would be irrelevant for the purposes of patentability that the organism or other material contains human DNA except indirectly insofar as certain DNA expresses features that are tied to personhood. (Hagen and Gittens 2008: 29)

European Patent Office

Directive 98/44/EC of the European Parliament and of the Council on the legal protection of biotechnological inventions indicates in Article 6 that:

1. Inventions shall be considered unpatentable where their commercial exploitation would be contrary to ordre public or morality ...

2. On the basis of paragraph 1, the following, in particular, shall be considered unpatentable:

(a) processes for cloning human beings;
(b) processes for modifying the germ line genetic identity of human beings;
(c) uses of human embryos for industrial or commercial purposes;
(d) processes for modifying the genetic identity of animals which are likely to cause them suffering without any substantial medical benefit to man or animal, and also animals resulting from such processes.

Also excluded from patentability are processes to produce chimeras from germ cells or totipotent cells of humans and animals (EU Dir. 98/44/EC, rec. 38).

Most international conventions and directives concern either human beings or nonhuman animals and do not consider the possibility of creatures of combined human and nonhuman origin. One exception to this rule is given in the guidelines of the European Patent Office. These guidelines consider both the genetic modification of human beings and the genetic modification of nonhuman animals. They also explicitly consider chimeras.

Nevertheless, even in a context where human-nonhuman animal combinations are considered, the guidelines are framed in a way that begs the essential question. The provisions of 6.2 (d) effectively allow the patenting of processes which modify the genetic identity of animals under certain conditions. In context 'animals' clearly refers to nonhuman animals (in contrast to human beings which are covered by 6.2 (b)). However, in the case of human-nonhuman combinations it may not be easy to say what counts as nonhuman, and thus whether a process should fall under (b) or under (d). The provisions of Directive 98/44/EC, rec. 38 exclude the creation of chimeras but it is not clear that the further genetic modification of a chimera or of a hybrid could not be patented under the provisions of the guidelines. For this issue to be clarified there would need to be a definition of a human being and nonhuman animal that was sensitive to the possibility of human-nonhuman combinations. (For a more extensive discussion of the European patenting guidelines relating to animal-human combinations, see Hagen and Gittens 2008.)

In November 2008, the European Patent Office finally refused a patent to the Wisconsin Alumni Research Foundation (WARF) for a stem cell technology that involved the destruction of human embryos (EP0 770 125, see N.R.D. 2009). The Office rejected the argument of WARF that the prohibition on patents only related to embryos subsequent to 14 days development. The Office referred to German and United Kingdom legislation, both of which define the human embryo as an embryo from, at least, the two cell stage. However, there is no similar consensus among European nations as to whether or when a human-nonhuman combination may be classified as human. It is therefore unclear how the European Patent Office would decide if a case were brought to patent a technique involving the destruction of human-nonhuman hybrid embryos.

United States Patent and Trademark Office

In 1997, the biologist Prof Stuart Newman in collaboration with science and technology commentator Jeremy Rifkin, filed a patent on a technique for combining human embryo cells with cells from the embryo of a monkey, ape or other animal to create a blend of the two (a chimera), in order (1) to keep others from pursuing such work for 20 years, or (2) for the US Patent and Trademark Office to reject it, effectively accomplishing the same thing. Eventually, in 2005, the US Patent and Trademark Office dismissed the claim, saying the chimera would be too closely related to a human to be patentable (Wiess 2005). It also indicated that it cannot define the proportion of human DNA in an organism for it to consider this as human (Hagen and Gittens 2008). More generally, it emphasized that the application did not meet the appropriate utility requirement in the face of morality concerns.

However, in 1999, the Federal Circuit Court of Appeals eventually rejected the concept of morality as a factor to be used to refuse a patent in the USA. But the Patent and Trademark Office has not abandoned its position that the possibility of patenting human beings would be contrary to public policy (Stoltzfus Jost 2009: 765). It also reiterated its position that human beings are not patentable matter (see application No. 10/308. 135, decision mailed 8/11/2004 at 22). This was probably to clarify that if certain human-nonhuman combinations were to be considered as 'human beings' then they could not be patentable.

PART TWO

Developments in the creation of human-nonhuman combinations

This part will examine the different procedures related to the creation of human-nonhuman interspecies entities. Examples of each practice will clarify exactly what is at stake, as well as demonstrating the relevance of the discussion for the twenty-first century by presenting three components.

First, the actual procedure will be clearly described in lay language without compromising scientific accuracy.

Secondly, the UK legal framework relevant to this procedure is examined, taking this as a useful example of the regulatory context in a large Western country which has recently enacted legislation in the area. In this regard, it should be noted that the UK Human Fertilisation and Embryology Authority which is mentioned in the legislation is currently under review as part of a rationalization of 'Arm's Length Bodies' and may eventually be replaced by another but similar regulatory body.

Finally, each short chapter will identify the key ethical issues that have been or that may be raised in relation to the procedure in question. These are the issues that will then be explored in the third part of the book.

3

Human-nonhuman transgenesis

For many years, scientists have been creating transgenic bacteria and animals in which some foreign (human or nonhuman) genes are deliberately inserted into the genome of living beings.

At first, researchers were very much concerned with the potential risks involved in creating transgenic organisms. This resulted in the 1975 Asilomar conference in California (USA) in which 140 researchers laid down procedures to organize safe and responsible transgenic experiments. Accordingly, the research was initially undertaken under very restrictive conditions of containment and safety. On the basis of experience, however, researchers then learned to distinguish between dangerous and safe genetic modifications which did not require a high level of containment. At present, the genetic modification of micro-organisms has become standard practice in academia and pharmaceutical companies (Van Steendam et al., 2006: 753).

As a result of this research, human insulin was the first medication produced using modern genetic engineering techniques. In this procedure, actual human DNA was inserted into a host cell (*E. coli* in this case). The host cells were then allowed to grow and reproduce normally, and because of the inserted human DNA, produced actual human insulin.

With respect to mammals, including humans, it is estimated that their genomes are made up of around 25,000 genes with the difference in species only representing a small variation in genes. Indeed, maybe only 100–1000 genes may be instrumental in defining a specific species with respect to behaviour and appearance. However, it remains unclear which genes are really necessary to create an entity which could be characterized as human. But the addition of a very small number of human genes to a nonhuman mammal is very unlikely to result in any significant difference from a species perspective (Bader 2009: 473).

In animals, the production of human growth hormone in the serum of transgenic mice, in 1982, was the first successful example that transgenic animals could produce human products for therapeutic use (Palmiter et al. 1982). Since then this technique has been successfully used to produce a variety of human therapeutic proteins in the milk, blood serum, urine and semen of mice, rabbits, sheep, goats and pigs (Primrose et al. 2001: 283–4). Transgenic chickens are now also able to synthesize human proteins in the 'white' of their eggs which scientists believe may eventually prove to be a valuable source of proteins for human therapy. In addition, mice with human immune-system cells and organ-donor pigs with human genes have been created which can pass on the human genes to subsequent generations (Logan and Sharma 1999).

It is generally very difficult to predict the consequences on animal welfare that the addition of one or a combination of genes may have on the animals but a number of experiments have shown that some may be compromised in their wellbeing (Nuffield Council on Bioethics, 2005: 80). A number of transgenic animals have suffered unexpected side effects such as those experienced by the 'Beltsville pigs' (named after the US Department of Agriculture research station where they were born). In these pigs, scientists introduced a human growth gene which was intended to make the pigs grow faster and produce less fat. But unfortunately, they began to demonstrate unforeseen symptoms, including diarrhoea, mammary development in males, lethargy and loss of libido as well as skin and eye problems. The pigs were in constant pain and had joints so diseased with chronic arthritis that when they tried to walk, they could only crawl on their knees (Mench 1999).

In addition, several technical obstacles have limited the amount of human genes that can be expressed in animals such as mice. Only a few human genes at a time can be successfully inserted into the mouse genome without interrupting essential mouse gene functions or creating a fatal combination. The procedure is in fact extremely inefficient with large numbers of animals being created in order to produce a single genetically modified strain. This means that the majority of animals that are produced do not express the desired genetic trait and are then euthanized (Nuffield Council on Bioethics, 2005: 99).

Specific ethical issues

In some quarters, concern has been expressed about the 'humanization' of animals through transgenesis. Indeed, the genetically modified sheep

producing pharmaceutical proteins or the pigs intended as sources of transplants, may technically be considered as 'humanized', in that they produce
'human' proteins rather than ovine or porcine equivalents. Nevertheless,
while these animals produce specifically human proteins, they do not
generally raise concern in relation to the display of characteristics generally
thought of as essentially human. It is also true that speaking of 'human
genes' may be misleading. Indeed, the insertion or excision of DNA in order
to replicate human genetic sequences may not involve any actual transfer of
human material, any more than they encourages the expression of significantly human features (Animal Procedures Committee 2001: 18–20).

In the light of the 'precautionary principle' such experiments could be
considered as having to proceed with extreme caution, if at all. On the other
hand this may seem unreasonably to impede progress not only towards
agricultural and medical benefit to human beings, but also on developments
that might improve animal welfare (for example, by increasing resistance to
disease or obviating the need for antibiotics or other invasive procedures).

The application of these principles cannot be resolved without further
analysis of the principles themselves. If this is not done the danger is that
ethical principles will not provide sufficient guidance and the gap will be filled
by prejudice or commercial interest.

4

Human-nonhuman gestation

Placing a human or human-nonhuman interspecies embryo in a nonhuman animal

The placing of a human embryo into an animal was addressed by the President's Council on Bioethics of the USA in its report entitled 'Reproduction and Responsibility: The Regulation of New Biotechnologies' published in 2004. In this document it is suggested that a bright line should be drawn at the insertion of *ex vivo* human embryos into the bodies of animals. Thus, an *ex vivo* human embryo entering a uterus only belongs to a human uterus. This prohibition would also apply to human-nonhuman embryos who have a few animal genes / cells.

In the influential UK 1984 Warnock Report, serious concern was also expressed in relation to any future attempt to place a human embryo into the uterus of a nonhuman animal for gestation. Because of this, the report went on to suggest that any placing of a human embryo into the uterus of a nonhuman species should be prohibited (Warnock 1984: 72).

In the report prepared in 2005 by the UK House of Commons Science and Technology Committee entitled 'Human Reproductive Technologies and the Law' it was moreover indicated that the placing of a human embryo in an animal raises difficult ethical issues. These include the special status of the embryo and animal welfare. However, this committee went on to note that once an embryo had been created but was not required for treatment, it must either be destroyed or used for research. In this regard, the House of Commons committee suggested that it could be argued that its special status demanded that it be used for potentially valuable research. In other words, the committee suggested that if a spare human embryo did become available for research then, because of the respect due to the embryo, researchers should ensure that it is used for the best possible ends. This implies that a human embryo could ethically be placed into a nonhuman animal if, for example, such a procedure could give rise to a better understanding of the causes of disorders such as infertility and miscarriage. The committee acknowledged, however, that it was unaware of any interest by scientists, at present, in undertaking such experiments (HCSTC 2005: 30–2).

In this respect, the House of Commons Science and Technology Committee indicated under a section entitled 'Placing a human embryo in an animal' that,

> the ethical problems concerning the use of embryos surplus to treatment are not clear cut, particularly if no embryo could be incubated in the animal for longer than the statutory maximum duration for *in vitro* culture. In considering the subject comprehensively we should not shy away from addressing difficult subjects which may widely be considered "taboo".

Nevertheless, the UK Human Fertilisation and Embryology Act 1990 was quite clear in its opposition to such a procedure. Section 3(3)(b) indicated that 'A licence cannot authorise placing a [human] embryo in any animal'.

This was re-emphasized when the Human Fertilisation and Embryology Act 2008 amended the 1990 Act to indicate in Section 4A (4) that 'A licence cannot authorise placing a human admixed embryo in an animal'.

However, there is apparently no complete prohibition in UK law with respect to placing human sperm and eggs together in the uterus of another animal, though, because gametes would need to be stored in order to undertake this experiment, a licence would be required for the storage (Lee and Morgan 2001: 87).

It should also be noted that the placement of any kind of embryo into the uterus of a protected animal in the UK, is automatically a regulated procedure for the purpose of the Animals (Scientific Procedures) Act 1986 if it is undertaken for an experimental or other scientific purpose whether or not the pregnancy is subsequently terminated (House of Lords, *Hansard*, Written Answers, 25 Jun 2007, Column WA103). This is the case regardless of whether the implantation required a surgical procedure – not only because of the nature of the intervention required to implant the material, but also because of the possible welfare problems that might arise post-implantation (personal communication to Dr C. MacKellar from the UK Home Office, 20 April 2009).

Experiments already undertaken

The mixing of human eggs and sperm in a sheep

In 1984, an experiment was reported in Australia in which researchers intro-duced human eggs and sperm into the fallopian tube of a sheep. However, no development was observed (Lee and Morgan 2001: 87; Wood and Westmore 1984).

Specific ethical issues

It is likely that this procedure would create grave and complex ethical diffi-culties in the manner in which the human character of the human embryos would be considered. There is indeed a risk that human embryos and human foetuses could be considered in the same manner as those of animals (with no special protection being granted) if they were to be found in an animal. As a result, this may further undermine the conferring of any respect and dignity on human embryos and foetuses. Clearly, in the unlikely event that a child were born as a result of such an experiment, there would be grave danger of physical and psychological harm to the child. On the other hand, if the pregnancy were deliberately terminated as part of the experiment, this would seem a further offence against the child. The example of xeno-gestation thus raises the issue of whether there are actions that are inherently incompatible with respect for a human embryo, even for one that is 'surplus' and has little prospect of future life.

In the light of the proportionality principle, there is widespread consensus that such experiments should be proscribed. There are no evident scientific benefits arising from such a procedure and there is no virtue in relaxing the current prohibition. Nevertheless, the radical stance taken by the House of Commons Science and Technology Committee shows the need for a clearer rationale for ethical judgments in this area, for indeed it cannot be guaranteed that in the future someone will not suggest a possible scientific benefit that could be derived from such experiments.

Placing human sperm into a nonhuman animal

Many national legislations, including the UK Human Fertilisation and Embryology Acts 1990 & 2008 do not specifically mention the placing of human sperm into an animal.

This was deliberate since a proposed amendment to the Human Fertilisation and Embryology Act 2008 which would have prevented human gametes being placed into animals was specifically voted down (by 308 to 183) by Members of Parliament in a UK House of Commons vote in October 2008 (*Hansard*, 22 October 2008, Division 287).

However, it is probable that one of the reasons (amongst others) why many countries prohibit bestiality (sexual activity between humans and nonhuman animals) is related to the deep revulsion towards the creation of possible human-nonhuman combinations through the biological placement of human sperm into an animal (Williams 1958). For example, in England, Wales and Northern Ireland, the Sexual Offences Act 2003 indicates under Section 69 (Intercourse with an animal), paragraph (1) that a man commits an offence if he has sexual intercourse with an animal. Under Scots common law, there is an offence of bestiality defined as 'unnatural carnal connection with a beast.' The last prosecution in Scotland took place in 1994 (personal communication from the Justice Directorate of the Scottish Government, 10 January 2008).

Experiments already undertaken

Chimpanzee-human hybrid entities

In 1926 the Russian scientist, Prof Ilya Ivanov, was sent to West Africa by the Communist regime to conduct experiments in which human sperm was implanted into female chimpanzees. Prof Ivanov had already established his reputation under the Tsar through the creation, in 1901, of the world's first centre for the artificial insemination of racehorses. The human-chimpanzee experiments, however, remained unsuccessful (Rossiianov 2002, Stephen and Hall 2005).

Specific ethical issues

Once more it is probable that the placing of human sperm into an animal creates very grave and complex ethical difficulties with respect to the conferring of moral status. Again such experiments could be considered as deserving to be proscribed on the basis of proportionality, because of the lack of any countervailing case. However, this issue also raises the question of whether there are any inherent ethical principles at stake in such cases.

Placing a nonhuman or a human-nonhuman interspecies embryo into a human

In some countries, such as in the UK, legislation exists prohibiting the placing of a nonhuman or human-nonhuman interspecies embryo in a woman (HCSTC 2005: 30–2).

Thus, the UK Human Fertilisation and Embryology Act 1990 indicated in Section 3(2)(a):

> No person shall place in a woman a live embryo other than a human embryo.

Moreover the Human Fertilisation and Embryology Act 2008 amended the 1990 Act to re-emphasize this prohibition in:

> Section 3(2)(a): No person shall place in a woman an embryo other than a permitted embryo.

> Section 4A(1): No person shall place in a woman:

> a human admixed embryo,
> any other embryo that is not a human embryo.

Specific ethical issues

Serious concerns exist that placing a nonhuman embryo into a woman would create grave and complex ethical difficulties. In the light of the proportionality principle, such experiments could be considered as deserving to be proscribed as there do not seem to be any useful scientific benefits arising from such a procedure and there is no virtue in relaxing the current prohibition. This case also raises the question of the abuse of women, an issue which, together with racism, was evident in experiments proposed in the early twentieth century. These were considered without seeking consent, but even with consent and even were safety issues to be addressed, placing a nonhuman animal inside a woman arguably has implications for the status of women and the portrayal of pregnancy. The human rights of women and the ethics of procreation thus join other principles, not least the status of the human embryo and the dignity of human nature, in demanding proper consideration. Again the issue is far more complex than balancing alleged scientific benefits against 'taboo'.

Placing nonhuman sperm into a woman

Again, in a number of countries, the placing of nonhuman sperm into a woman is strongly condemned.

For example, in the UK Human Fertilisation and Embryology Act 1990 Section 3 (2)(a) reads:

> No person shall place in a woman any live gametes other than human gametes.

This remained the case when Human Fertilisation and Embryology Act 2008 amended the 1990 Act in Section 3 (2)(b):

> No person shall place in a woman any gametes other than permitted eggs or permitted sperm.

And 4A (1)(c):

> No person shall place in a woman any gametes other than human gametes.

Moreover, as argued above, one of the reasons amongst others, why many countries prohibit human-animal sexual activity is related to the possibility that a human-nonhuman creature might be conceived through the biological placement of animal sperm into a woman. For example, in England, Wales and Northern Ireland, the Sexual Offences Act 2003 indicates under Section 69 (Intercourse with an animal), paragraph (2) that a woman commits an offence if she has sexual intercourse with an animal. Both men and women would be guilty under the same offence.

Under Scots common law, there is an offence of bestiality defined as 'unnatural carnal connection with a beast'. The last prosecution in Scotland took place in 1994 (personal communication from the Justice Directorate of the Scottish Government, 10 January 2008). This is relevant both to placing human sperm in a nonhuman animal and to placing nonhuman sperm into a woman.

Specific ethical Issues

The placing of nonhuman sperm into a woman would create the same grave and complex ethical difficulties as outlined above in relation to placing a nonhuman or admixed embryo into a woman.

5

Human-nonhuman hybrid embryos

Embryos containing cells made up of both human and nonhuman chromosomes

Nonhuman eggs into which human nuclei are inserted

The insertion of human nuclei into nonhuman eggs has already led to some concern in the press (Henderson 2004). This is because scientists from the Monash Institute of Reproduction and Development in Australia, have demonstrated that any complete set of chromosomes could, theoretically, be considered as gametes if they were introduced into an egg (Highfield 2001; Lacham-Kaplan, Daniels and Trounson 2001). The researchers found a way of 'fertilizing' non-enucleated mouse oocytes by injecting somatic cell nuclei taken from adult male mice. And following chemical activation of the 'fertilized' oocytes and the extrusion of two secondary polar bodies, embryos could be formed containing two sets of chromosomes which could further develop. Thus, if the nuclei of adult human cells were inserted into nonhuman eggs, concern may be expressed that the human nuclei could, theoretically, 'fertilised' the nonhuman eggs.

Experiments already undertaken

Frog-human hybrid entities

In 2003, a team of scientists at Cambridge University fused the nuclei of adult human cells with immature frog (Xenopus) oocytes, with the ultimate aim of producing rejuvenated master cells that could be grown into replacement tissues for treating disease. In this experiment some kind of very early biological development was initiated, the extent of which is difficult to ascertain from an ethical perspective (Henderson 2004; Byrne et al. 2003).

Regulation in place

In the UK, the Human Fertilisation and Embryology Act 2008 amends the 1990 Act of the same name to indicate that no person shall bring about the creation of a human admixed embryo except in pursuance of a licence. In this regard, a human admixed embryo includes any embryo which contains nuclear or mitochondrial DNA of a human *and* nuclear or mitochondrial DNA of a nonhuman animal, but in which the nonhuman DNA is not predominant.

Specific ethical Issues

It is possible to question why human donor nuclei were used in the Cambridge experiment rather than those from less ethically sensitive beings, such as other mammalian species in order to perform such basic research aimed at better understanding the mechanism of nuclear reprogramming. This again relates to proportionality.

In addition, from an ethical perspective there is reason to explore the under-lying rationale for the response made by the UK Government, (Department of Health 2000b), to the Recommendations made by the Chief Medical Officer's Expert Group entitled 'Stem Cell Research: Medical Progress with Responsibility' (also called the 'Donaldson Report', Department of Health, 2000a). The response stated:

> The mixing of human adult (somatic) cells with the live eggs of any animal species should not be permitted.

Human-nonhuman chromosome transplant

The possibility of transplanting chromosomes between animals and human beings would create new ethical problems with respect to the manner and extent in which the new beings should be considered and whether they should have animal or even human rights. Human beings normally contain 46 chromosomes of which 23 originate from each parent. A question can then be raised concerning the manner in which an interspecies chromosomal transplantation would affect the manner in which one considers a creature. Should a living being be considered to belong to a specific species if the majority of its chromosomes belong to this species? A further question can be considered if a living being was created with chromosomes originating from several different species.

Regulation in place

In the UK, the Human Fertilisation and Embryology Act 1990 did not include any provisions which could specifically address human-nonhuman chromosome transplants. However, the Human Fertilisation and Embryology Act 2008 which amends the 1990 Act sought to address this legal vacuum by indicating that no person shall bring about the creation of a human admixed embryo except in pursuance of a licence. In this regard, a human admixed embryo includes any embryo which contains nuclear or mitochondrial DNA of a human *and* nuclear or mitochondrial DNA of a nonhuman animal, but in which the nonhuman DNA is not predominant.

Experiments already undertaken

Mouse-human hybrids

In 1989, American researchers reported the introduction of human chromosome fragments into fertilized mouse eggs which they could then follow as the embryos grew and developed into mice. The scientists indicated that human chromosomes were detected in the tail of at least one mouse (Science Watch 1989).

In 2005, a team of UK scientists was also able to successfully transplant a human chromosome into mice. To create these mice, the team first extracted chromosome 21 from human cells and sprayed them onto beds of stem cells taken from mouse embryos. Any stem cells that absorbed human chromosome 21 were injected into three-day-old mouse embryos which were then re-implanted into their mothers. The newly born mice carried copies of the chromosome and were able to pass it on to their own young which contained the human chromosomes in their cells (Sample 2005; O'Doherty et al. 2005; Shinohara et al. 2001).

One should note that this work was conducted in order to provide a better mouse model for Down's syndrome, which may lead to positive clinical benefits for affected individuals. If these particular mice were considered either to be substantively more human or to be undesirable entities as a result, then it may be necessary to carefully consider the possible implications of these conclusions for members of the population with Down's syndrome who also have an additional chromosome 21.

Specific ethical issues

It is possible to be very concerned that this procedure would create grave and complex ethical difficulties if mouse-human embryos were produced containing cells made up of both mouse and human chromosomes with thousands of human genes (Bader 2009: 473). Here there is a more positive prospective benefit, but this benefit should be set against the context of the current practice of eugenic termination of pregnancy for Down's syndrome. This raises the issue of disability and whether such forms of research enhance or undermine respect for persons with disability. Apart from this issue, similar considerations apply in this case as with transgenic combinations, i.e. the application of conflicting ethical principles cannot be resolved without further analysis of the principles themselves.

Mixing of human and nonhuman gametes (true hybrids)

Full interspecies genetic hybrids involve the creation of an embryo using gametes from different species. The latter will usually only be possible between closely related species, and even then the hybrids are likely to be sterile. Interestingly, natural intercourse between different kinds of animals has fascinated humanity ever since it was observed. Aristotle in his book *Generation of Animals* (Book 2.7) described at some length the results of such unions. Typical examples of natural true hybrids are ligers which are the offspring of a male lion and a female tiger, though the creation of such animals has now been stopped in India on the basis of animal welfare. True hybrids between sheep and goats have also been reported (Bonnicksen 2009: 61). A more common animal is the mule which is the result of a male donkey mating with a female horse. In these examples, the species are similar enough to enable viable offspring to be produced though they are usually infertile themselves. Early embryonic true hybrids between more different species may be obtained but would necessitate the artificial combination of the gametes in the laboratory.

When natural hybridization takes place, mitochondria are usually inherited from the oocyte, as there is a special mechanism that eliminates mitochondria brought into the early embryo by sperm.

However, in cross-species hybrids, this mechanism may not always operate resulting in embryos in which the mitochondria comes from both the oocyte and the sperm (Kaneda et al., 1995).

It is unknown whether any species of animal's gametes could successfully undergo hybridization with a human gamete. The closer an animal is to humans, in biological terms, the more likely there would be success. With respect to a successful hybridization between a human and a chimpanzee the biologist Richard Dawkins conceded that 'It cannot be ruled out as impossible, but it would be surprising' (Rabderson and Dawkins 2009). Humans are significantly more advanced than any other species of animal, making the chances of successful hybridization with any nonhuman species low, no matter how closely related they are to humans (Lord Darzi of Denham, *Lords Hansard*, 19 Jun 2008, Column WA188) though it should be noted that human sperm has been reported to penetrate gibbon eggs (Bedford 1977).

In this context, some have already defended the possibility of fertilizing nonhuman eggs with human sperm for the testing of *in vitro* derived gametes which have also been called 'artificial gametes'. These stem cell derived gametes could then be used in the treatment of many patients who have survived cancer and cannot create their own gametes (Dr E. Harris, House of Commons Debates, *Hansard*, 19 May 2008, Column 55).

It has also been claimed that cytoplasmic sperm injection to generate true hybrids to be grown beyond the two-cell stage could be of use in the understanding of serious mitochondrial diseases (Lord Darzi of Denham, *Lords Hansard*, 19 Jun 2008: Column WA188).

The mixing of nonhuman and human gametes to form human-nonhuman entities is possible in a number of countries, such as the UK (Casabona and Beiain 2009), where the fertilization of animal eggs with human sperm can be used to determine the quality of this sperm, subject to obtaining a licence from the Human Fertilisation and Embryology Authority.

In the UK, the Human Fertilisation and Embryology Act 1990 indicated in Article 4. (1)(c) that:

No person shall mix gametes with the live gametes of any animal, except in pursuance of a licence.

But, as indicated under Schedule 2, Article 1. (1)(f) of this Act, a licence could authorize, in the course of providing treatment services, 'mixing sperm with the egg of a hamster, or other animal specified in directions, for the purpose of testing the fertility or normality of the sperm, but only where anything which forms is destroyed when the test is complete and, in any event, not later than the two cell stage'.

However, the UK Human Fertilisation and Embryology Act 2008 amended the 1990 Act to indicate that no person shall bring about the creation of a human admixed embryo except in pursuance of a licence. In this regard, a human admixed embryo includes any embryo created by using human and nonhuman gametes.

Moreover, the new Act amended the 1990 Act to specify in Section 4A (3) that:

A licence cannot authorize keeping or using a human admixed embryo after the earliest of the following:

(a) the appearance of the primitive streak, or
(b) the end of the period of 14 days beginning with the day on which the process of creating the human admixed embryo began, but not counting any time during which the human admixed embryo is stored.

Experiments already undertaken

Genetic human-hamster hybrid embryos

The hamster test is a well-established test in the UK used to examine the motility and normality of sperm.[1] Whereas, the zona pellucida[2] of the ovum and the tip of the sperm are generally species specific, very closely related species such as the horse and the donkey can interbreed. However, the hamster is the only known exception to possess a removable zona pellucida which would otherwise repel the sperm from another species.

Thus, the hamster zona is removed following a specific biochemical treatment in order to undertake what is known as a zona free hamster oocyte penetration test which generally requires the mixing of 40 golden hamster eggs with a man's sperm. After being incubated for three hours or more at 37°C, a judgement is made relating to the percentage of eggs that have been penetrated by the sperm, on the basis of which a conclusion is made as to the fertility of this sperm. If the test is only completed after 24 hours have elapsed following fusion, the embryo would normally have divided into two

---

[1] 'This is variously known as 'Hamster Egg Penetration Test' (HEPT), the 'Hamster Zona-Free Ovum test' (HZFO test), or the 'Sperm Penetration Assay' (SPA). It is a test of the ability of a man's sperm to penetrate a hamster egg that has been stripped of its outer layer, the 'zona pellucida'. The test focuses on the ability of the sperms to begin the process of fertilization by progressing through the membrane of the egg, in scientific terminology it tests the sperms' ability to 'undergo capacitation and the acrosome reaction'.

[2] The zona pellucida is a membrane surrounding an ovum and is vital for its survival and enabling fertilization. In humans, five days after the fertilization the membrane degenerates and decomposes to be replaced by the underlying layer of trophoblastic cells.

cells at which stage it is destroyed. The importance of the test is its value in studying the biological constitution of human sperm, and hence the male contribution to genetic abnormalities and infertility, thought to affect one in 16 of the male population (Lee and Morgan 2001: 89–90).

At the time when the Human Fertilisation and Embryology Act 1990 was put through parliament in the UK, the 'Hamster Egg Penetration Test' (HEPT) in which human sperm is mixed with a hamster's egg, was one of the few valid tests available to measure the viability of some patients' sperm. However, the introduction of Intra Cytoplasmic Sperm Injection (ICSI) and other treatments has now made HEPT effectively obsolete for testing sperm prior to treatment. A similar technique has also sometimes been used in the UK for research into the viability of sperm. But at the time of writing, the UK Human Fertilisation and Embryology Authority has not given any licences for HEPT to any treatment or research centres since 2003.

Although currently there are no research projects licensed for HEPT in the UK, the case for use of the technique is still legally valid. Any research project which intends to use the technique would need to prove that it had considered the ethical implications and that the use of HEPT was necessary or desirable under the purposes for which the HFEA can issue research licences.

Specific ethical issues

This is the most extreme form of mixing of human and nonhuman and raises, in the most acute form, the question of whether there is any inherent ethical problem generating creatures whose humanity is ambiguous. It is very probable that this procedure would create grave and complex ethical difficulties, especially if it gave rise to a living embryo without any clear moral status.

In this regard, the Political Scientist Andrea Bonnicksen indicated that the creation of a human-nonhuman primate through hybridization 'would violate the core bioethical principle of nonmaleficence because it would bring both physiological and psychological harms. No imaginable benefit would accrue except existence, and even that would be hard to justify' (Bonnicksen 2009: 69).

6

Human-nonhuman cytoplasmic hybrids (cybrids)

With regard to 'cloned' embryos created with donor nuclei and recipient enucleated oocytes from different species, the nomenclature used to describe such novel entities has been varied. The term hybrid may be misleading if one is normally used to restricting the term to the progeny resulting from sexual reproduction, whilst the term chimera may be misleading if one normally only associates this with an organism containing distinct population of cells which differ in their genetic origin.

Though some recent legislation, such as in Australia (Prohibition of Human Cloning for Reproduction and the Regulation of Human Embryo Research Amendment Act 2006) has brought the creation of human-nonhuman embryos through Somatic Cell Nuclear Transfer (SCNT) under the category of hybrids, because such entities would share characteristics with both hybrids and chimeras it may be appropriate to address these combinations in a separate paragraph under cybrids (cytoplasmic hybrids). However, this does not mean that these entities are not hybrids or chimeras but only that they share features from both groups.

Developments in the creation of interspecies cytoplasmic hybrids

The use of interspecies nuclear transfer was first mentioned in 1886 during an investigation of the roles of the nucleus and cytoplasm in heredity (Rauber 1886). This attempt involved the transfer of zygotic nuclei between frog and toad eggs, which resulted in the development arrest of both eggs.

A number of further experiments were published relating to mammalian nonhuman interspecies embryos which have been reported to develop to the blastocyst stage (50–100 cell embryo). These include:

> Horse donor cell and cow oocyte (Tecirlioglu, Guo and Trounson 2007),
> Monkey donor cell and rabbit oocyte (Yang et al. 2003),
> Various mammalian species and cow oocyte (Dominko 1999),
> Mountain bongo antelope donor cell and cow oocyte (Lee et al. 2003),
> Buffalo donor cell and cow oocyte (Lu et al. 2005),
> Dog donor cell and yak oocyte (Murakami et al. 2005),
> Panda or cat donor cell and rabbit oocyte (Wen et al. 2005),
> Takin donor cell and yak or cow oocyte (Li et al. 2006).

Other examples of interspecies embryos were also reported but these ceased developing at stages earlier than the blastocyst:

> Antarctic minke whale donor cell and cow or pig oocyte (Ikumi et al. 2004),
> Rabbit donor cell and cow oocyte (Shi et al. 2003).

The procedure has been used, as well, to attempt to clone endangered species where oocytes of a specific species were not readily available.

A few studies have also demonstrated the possibility of obtaining embryonic stem cell lines from interspecies cytoplasmic hybrid embryos. In this way, mouse embryonic stem cell lines were derived from embryos created from mouse somatic cells and bovine oocytes in 2006. These cells differentiated into various typical embryonic germ tissue types and contributed to chimeric offspring when transferred to mouse blastocysts (Vogel 2006).

The first development of human-nonhuman cytoplasmic hybrids was reported in the 1970s in the creation of mouse-human cytoplasmic hybrids, using two somatic cells. At the time, the experiment was undertaken to investigate the interactions between nuclear and mitochondrial genomes (Clayton et al. 1971; De Francesco et al. 1980; Giles et al. 1980).

Objectives in the creation of interspecies cytoplasmic hybrids

It has been suggested that the use of nonhuman oocytes stripped of their chromosomes as recipients of human somatic nuclei with the aim of generating human embryonic stem cell lines without the need for human oocytes, may constitute a solution to the problem of limited supplies of human oocytes (National Academy of Sciences 2005: 34). In the future, however, researchers may, conceivably, be able to obtain large numbers of human egg-like cells derived from human pluripotent cells (McLaren 2007: 339). These human stem cells lines could allegedly then be developed into new treatments for neurological disorders such as Alzheimer's and Parkinson's diseases (Minger 2006).

This position was emphasized by the UK House of Commons Science and Technology Committee in its 2007 report entitled 'Government Proposals for the Regulation of Hybrid and Chimera Embryos' in which the Members of Parliament indicated their belief that the

> use of animal eggs in the creation of cytoplasmic hybrid embryos will help to overcome the current shortage of human eggs available for research and that use of animal eggs is required to enable researchers to develop the practical techniques which may be required for eventual production of cell-based therapy through this method using human eggs. (HCSTC 2007: 33–4)

It has also been proposed that these human-nonhuman cytoplasmic hybrids may be useful as a potential replacement for animal use in experiments, such as in testing the toxicity of new drugs, since animals do not always highlight potential problems. Thus, by using these stem cells produced by human-nonhuman cytoplasmic hybrids, high throughput drug screens could be designed to screen drug effects in specific human cell types, especially those that may be vulnerable to toxicity or those that form the drug's target tissue.

Another use of human-nonhuman cybrids, suggested by some scientists, rests in their potential to become valuable research tools to increase the understanding of the reprogramming process of somatic nuclei, which could be a long term solution to the problems of tissue rejections (National Academy of Sciences 2005: 34).

Challenges in the creation of interspecies cytoplasmic hybrids

Scientific commentators agree that a number of mechanisms need to be effective for cytoplasmic hybrid embryos to develop successfully (HFEA 2007b, paragraph 5.20). This includes surmounting the potential incompatibility which exists if the mitochondria in a cell come from a different species than the nucleus.

Mitochondria, which are part of the main power system for a cell, should be capable of replicating to form new mitochondria in dividing cells and expressing their proteins to enable cell survival and development. (For the number of mitochondria in mammalian eggs, see Reynier et al. 2001.) Moreover, the proteins encoded by the mitochondria must be able to interact with those of the nucleus in order for the cell to produce the energy it requires (HFEA 2007b, paragraph 4.4.4.6).

Researchers have shown, however, that when the biological distance between two animal species used to create cytoplasmic hybrids is increased, the interactions between the mitochondria and nuclei become increasingly dysfunctional (HFEA 2007b, paragraph 5.21).

Before the 14-day stage, the embryo is generally able to rely on proteins and genetic messages that were already present in the animal oocyte (HFEA 2006: 45) but with further development and even if some embryonic cells do survive, the embryo as a whole would not if there were not enough cells able to undergo the rapid cell divisions, cell movements and induction events (where one tissue type influences another) that characterize this phase of development (HCSTC 2007, HC272-II, Memorandum 61 Supplementary submission from Dr Robin Lovell Badge, 1).

Some scientists, on the other hand, have suggested that as cytoplasmic hybrid embryos develop towards the blastocyst stage (50–100 cells stage) the gene products, such as the proteins, will gradually become more human. In other words, that by 14 days the embryo would be entirely human with respect to protein and genetic material apart from 13 proteins encoded by the animal mitochondria (HFEA 2007b, paragraph 5.18). This assertion, however, has never been verified by experimentation.

Moreover, the suggestion made by some researchers that human-nonhuman cytoplasmic hybrid embryos, or any cells derived from them, could become functional if they were inserted with supplementary human mitochondria (HFEA 2007b, paragraph 5.24) still needs to be fully established.

It is also important to note that this situation would be similar to one already reported by scientists during the transfer of the donor nucleus, in that some cytoplasm containing donor mitochondria is also likely to be transferred into the recipient oocyte. However, because there are already varying degrees of interspecies incompatibility between mitochondria and nuclear functions, it cannot be ruled out that the biological difficulties related to using cells containing a mixture of nonhuman and human mitochondria and a human nucleus would be even more considerable. (For a review of how mitochondria may influence the outcome of interspecies cloning, for the timing of the nuclear genome being switched on in different species, and for coining of the term 'cybrid', see St John and Lovell-Badge 2007.)

Furthermore, not only would the amount of donor mitochondria transferred likely be affected by (1) the position and numbers of mitochondria in relation to the nucleus, (2) the type of donor somatic cell used and (3) the nuclear transfer technique employed (HFEA 2007b paragraph 4.3.2.8) but the possibility for human-nonhuman cytoplasmic hybrid embryos to somehow discard nonhuman mitochondria is improbable. It is far more likely that nonhuman mitochondria components will remain in human-nonhuman cytoplasmic hybrid cells and may even be selectively maintained.

Researchers have shown that in a cloned macaque-rabbit embryo, though the genetic material from the somatic cell mitochondria (which came with the nucleus of a macaque) and the recipient enucleated oocyte (from a rabbit) were both present from the one cell stage to the morula stage (16–32 cell embryo), the genetic material from the macaque mitochondria decreased dramatically at the blastocyst stage, 50–100 cells (Yang, C. X. et al. 2004). As a consequence, the interspecies dysfunction between the genetic material from the macaque nucleus and the rabbit mitochondria may possibly get worse with development.

Interestingly, some researchers have also reported that the reverse may also be true with genetic material coming from the donor mitochondria being seen to persist in the embryos up to the 16 cell stage or even blastocyst stage (HFEA 2007b paragraph 4.3.2.8.). A possible explanation for this observation is that the biological distance between species partners may affect the preservation of the genetic material from the donor mitochondria (HFEA 2007b paragraph 4.3.2.16).

In this context, it should be remembered that mitochondrial abnormalities have been shown to be a key factor in many neurodegenerative disorders such as Alzheimer's and Parkinson's diseases (for the role of mitochondria in human disease, see Wallace 1999). Because of this, using stem cells from dysfunctional human-nonhuman cytoplasmic hybrids to treat such neurodegenerative disorders would be liable to result in considerable medical risks. Similarly, the use of such stem cells to treat heart or liver complaints could create a number of difficulties owing to the high level of mitochondrial activity in these organs. Thus, very few proponents of cytoplasmic hybrids have advocated their use in clinical treatments but have, instead, merely proposed that they be used for *in vitro* research.

In summary, a lot more research is required with nonhuman interspecies cytoplasmic hybrids. In addition, it should be remembered that even when cloned animal embryos are produced with the nucleus and the oocyte originating from the same species, the somatic genome of the donor cell must be reprogrammed to allow the correct expression of genes for cell division and embryonic development. But difficulties relating to this well-known and well-described process account for the low efficiency of cloning in all species tested so far and for the now recognized defects and deficiencies in cloned animals. Thus, these problems are likely to only be magnified in cytoplasmic hybrids as different species may have different mechanisms for reprogramming the genetic material in their cells (HFEA 2007b paragraph 4.4.2.5). Indeed, many crucial genetic-protein interaction elements are found in the part of the nonhuman egg used for making cytoplasmic hybrids and these may be unable to correctly function with foreign genetic material from another species. Cytoplasmic hybrids, right from the beginning and throughout all later developmental stages, may be unnaturally dysfunctional.

Nonhuman eggs stripped of their chromosomes into which human nuclei are inserted

In the UK, the Human Fertilisation and Embryology Act 1990 did not have any provisions which could specifically address the creation of human-nonhuman cytoplasmic hybrids. However, the Human Fertilisation and Embryology Act 2008 which amends the 1990 Act sought to address this legal vacuum by indicating that no person shall bring about the creation of a human admixed embryo except in pursuance of a licence. In this regard, a human admixed embryo includes:

1 an embryo created by replacing the nucleus of an animal egg or of an animal cell, or two pronuclei, with (a) two human pronuclei, (b) one nucleus of a human gamete or of any other human cell, or (c) one human gamete or other human cell, and

2 any embryo which contains both nuclear or mitochondrial DNA of a human and nuclear or mitochondrial DNA of an animal but in which the animal DNA is not predominant.

Experiments already undertaken

Gametal cow-human hybrid embryos

The company Advanced Cell Technologies was reported, in November 1999, to have created the first human embryo clone using human-nonhuman cybrid technology. This was achieved when the nucleus of an adult human cell was inserted into a cow's egg stripped of its chromosomes in order to create a cloned embryo. This embryo was left to develop and divide for 12 days before being destroyed (BBC News Online 1998; BBC News Online 1999).

Similarly, Prof Panayiotis Zavos who runs a fertility laboratory in the USA claimed, in September 2003, to have created around 200 cow-human cybrid embryos that lived for around two weeks and grew to several 100 cells in size and beyond the stage at which cells showed the first signs of development into tissues and organs. It was also noted that they appeared to have normal human DNA (Coghlan 2003; Leake 2003). Moreover, his research team indicated, in experiments published in 2006, that human mitochondrial DNA was detected in almost all samples of interspecies cloned embryos, including morulae and blastocysts, which clearly indicated that mitochondria from the fused human donor cells were carried over to the enucleated bovine oocytes (Illmensee, Levanduski and Zavos 2006)

In April 2008, scientists in the UK also claimed that they had created human-nonhuman cytoplasmic hybrids by injecting the nuclei of human skin cells into eggs taken from cow ovaries which had virtually all their genetic material removed (BBC News Online 2008). The embryos survived for up to three days and were part of medical research into a range of illnesses. In an answer to a Parliamentary Question in the House of Lords, the UK government indicated that, according to the HFEA, 222 nonhuman eggs have been used, so far, in attempts to create human admixed embryos. (*Lords Hansard*, 23 Jun 2010: Column WA184).

In this context, however, a team of scientists led by Robert Lanza of Advanced Cell Technology in the USA, indicated in February 2009 that human-cow hybrid embryos failed to grow beyond the 16-cell stage. They also failed to properly express genes thought to be critical for the creation of pluripotent cells (Chung et al. 2009).

Gametal rabbit-human hybrid embryos

In August 2003, Hui Zhen Sheng of Shanghai Second Medical University, China, announced that rabbit-human cybrid embryos had been created by fusing adult human cells with rabbit eggs stripped of their chromosomes. Using donor cells from the foreskins of a five-year old boy and two men, and facial tissue from a woman, the researchers created rabbit-human hybrid embryos which developed to the approximately 50–100 cell stage that forms after about four days of development. Moreover, the scientists claimed that they were able to derive human embryonic stem cells from these embryos which were similar to those of conventional human embryonic stem cells, including the ability to undergo multilineage cellular differentiation (Abbot and Cyranoski, 2001; Sheng et al. 2003).

This research, nevertheless, has yet to be replicated by an independent group though Hui Zhen Sheng's team did indicate that they were able to make new human-bovine cybrid embryos in 2008 (Sheng et al. 2008). However, a team of scientists led by Robert Lanza of Advanced Cell Technology in the USA indicated in February 2009 that human-rabbit hybrid embryos failed to grow beyond the 16-cell stage. They also failed to properly express genes thought to be critical for the creation of pluripotent cells (Chung et al. 2009).

Specific ethical issues

Some of the most relevant questions relating to the ethical discussion concerning the creation of human-nonhuman cybrids arise from a scientific perspective and the possible usefulness of these entities. In this regard the following points need to be considered:

1. Experiment needs confirming

First, the scientific merit and validity of these experiments remain to be confirmed by an independent research group repeating the procedures. This is especially necessary following the seemingly disparate results having been published.

2. Difficulty in interpreting results

Regarding the use of human-nonhuman cybrids for research into diseases, it is noted that there are so many profound epigenetic flaws in cloned embryos (even using eggs of the same species), that to use embryos created by inter-species nuclear transfer would be liable to become a study of artefacts. In other words, it would be difficult to interpret the results since the potential usefulness of an experiment, to add to the sum of knowledge, is only as good as its capacity for interpretation. The biologist Professor Sir John Gurdon indicated that: 'Scientifically … I'm not persuaded it will work. If you put cells from one species into the egg of another, the egg may divide, but you could get a lot of genetic abnormality that won't lead to good quality stem cells' (Batty 2007).

3. Risk of creating new diseases

The creation of human-nonhuman cytoplasmic hybrids would also be associated with possible risks in the transmission of nonhuman diseases to humans or the creation of new diseases if any of the embryos were implanted into a host or not kept in a bio-secure environment.

For example, in cytoplasmic hybrids, both the mitochondria and the cytoplasm may contain retroviruses which could prove problematic (AMS 2007: 27). These retroviruses are RNA viruses which replicate in a host cell to produce DNA from their RNA genome. The DNA is then incorporated into the host's genome where the viruses are then replicated and multiplied. However, studies of mammalian mitochondrial DNA do not indicate any evidence, as yet, of endogenous retroviral genomes. If considered necessary, this could be further verified through studying the complete mitochondrial DNA sequences of nonhuman animals that might provide oocytes for cloning. If endogenous retroviral sequences cannot be found, this may address the risk of mitochondria transferring retroviruses on to humans through cytoplasmic hybrids (AMS 2007: 27).

On the other hand, the nuclear genomes of cows and rabbits do contain endogenous retroviral genomes (for instance, a replication-defective, endogenous beta-retrovirus of rabbits is described in Griffiths et al. 2002). It is therefore possible that the cytoplasm in the rabbit or bovine oocytes, which are emptied of their nuclei, may contain RNA transcripts corresponding to retroviruses which might conceivably re-integrate into the DNA of the transferred donor nucleus such as a human nucleus. Though this prospect is not impossible, scientific commentators believe that, on balance, this may be very improbable.

To establish whether such a possibility represents a real difficulty, expression profiles for endogenous retroviruses could be sought for oocytes from potential recipient species and if they are found, whether the retroviruses are replication-competent or whether they are harmless. Moreover, if the therapeutic use of cell lines derived from such cytoplasmic hybrids were to be envisaged as a real possibility, screening for such endogenous retroviruses could be considered (AMS 2007: 27).

Finally, it should be noted that the risk to laboratory workers or the wider public of endogenous retroviruses from cytoplasmic hybrids are unlikely to be any greater than that associated with regular cell culture procedures. There are already many human and mouse cell lines that release retroviruses, including endogenous retroviruses, capable of infecting human cells in regular use in a number of research facilities without any untoward consequences. However, safety concerns will necessitate further examination if cytoplasmic hybrid research generates products that may enventually be used in treatments (AMS 2007: 27).

4. Immunological incompatibility

From the perspective of immunological compatibility, it is doubtful whether the cells arising from human-nonhhuman cytoplasmic hybrids, in which a nonhuman egg is used, would become more human with development and could, therefore, be used in clinical applications. Since mitochondria have been shown to cause an immune reaction (Attari and Chomyn, 1995), it is not impossible that the presence of nonhuman mitochondria or other nonhuman elements in cells may also cause serious immune reactions in the host.

It should be noted that even when an animal is both the donor of the egg and the transplanted nucleus, there is a possibility of generating foreign cellular factors that will cause immune reactions. The first report of such an experiment conducted in mice, using embryonic stem cells from mouse-mouse homologous nuclear-transplant cloned embryos, showed that the stem cells were, in fact, not immune-privileged (Rideout et al. 2002). Because of this, and other factors, it is extremely unlikely that cells obtained from human-nonhuman cytoplasmic hybrids would be safe, from an immunological perspective, in human patients.

5. Impact on women

In addition to scientific considerations, a key element in the ethical evaluation of cybrid research is the impact of this research on women. Indeed the UK 2002 House of Lords report entitled 'Stem Cell Research' proposed that the placement of a human nucleus in an animal egg, as a way of creating human embryonic stem cells for research, could be seen as more acceptable than using embryos created by Cell Nucleus Replacement with human eggs, precisely because it would lessen the impact on egg donors (HL 2002, paragraph 8.18: 42).

The argument, however, that the use of cytoplasmic hybrids was a way of protecting women from exploitation (as cow eggs would be used) has been heavily criticized by the feminist scholar Françoise Baylis. It is clear that nonhuman eggs were used only to improve the efficiency of a technique which would then return to using women (Baylis 2008). This is reflected from the behaviour of the International Centre for Life in Newcastle, the only centre in the United Kingdom to do cytoplasmic hybrid research. At the same time as pursuing this research the same laboratory continued to pay women in kind for their eggs, providing half the cost of fertility treatment, equivalent to £1500. Such a practice is archetypal of the commodification of women's bodies criticized by a number of feminist commentators (Dickenson 2006; George 2008; Reynolds and Fogel 2009).

Conclusion

In the light of the preceding remarks, it is important to take seriously the UK Chief Medical Officer's Expert Group Report of June 2000 entitled 'Stem Cell Research: Medical Progress with Responsibility' (Donaldson Report) which indicated, in the context of using nonhuman eggs, as a shell to carry the human cell nucleus to produce stem cell lines, that: 'Most researchers active in this field do not regard this as a realistic or desirable way forward' (Department of Health 2000a, paragraph 2.34). In addition it found that 'the mixing of human adult (somatic) cells with the live eggs of any animal species should not be permitted'. Indeed, despite an intense media battle during the passage of the 2008 Human Fertilisation and Embryology Act, the scientific merit of this avenue of research remains highly doubtful. As the German National Ethics Council remarked in 2011, recent studies paint an increasingly critical picture of the development potential of human-nonhuman cybrids (Deutscher Ethikrat 2011: 26–7).

Furthermore, while cytoplasmic hybrids do not seem to raise the issue of perplexity to the same extent as true hybrids, it has been argued, both by opponents and by advocates of cybrid research, that there are similar questions of principle involved. These processes represent the mixing of human and nonhuman species. However, if cybrids do represent the crossing of a Rubicon, this provides a stronger reason for proscribing such research altogether as was recommended by the UK Government in August 2000 in its response to the Recommendations made by the Chief Medical Officer's Expert Group (Department of Health 2000b). Nevertheless, deeper reflection on the principles is required before it can be determined whether cybrids should indeed be placed together with true hybrids, or whether they might belong to a separate ethical category.

7

Human-nonhuman chimeras

Chimeras, unlike genetic hybrids, consist of combinations of cells, tissue and organs from two different biological entities whether or not they are of the same species. Unlike the situation in hybrids, there is no mixing of genetic material inside the individual cells of a chimera. It should also be noted, that every human person is a kind of chimera since most of the cells in a human person are not his or her own, nor are they even human. Indeed, more than 500 different species of bacteria exist in human bodies, making up more than 100 trillion cells. Because human bodies are made of only some several trillion human cells, they are, therefore, somewhat outnumbered by the aliens. It follows that most of the genes in a human person are actually bacterial (Nicholson et al. 2004).

From a historical perspective, a number of human-nonhuman chimeras with human blood, neurons, germ cells, and other tissues have already been created a few decades ago (Behringer 2007). However, it was generally in the animal kingdom that such chimeras have been studied. For example, the 'geep' was a combination of goat and sheep cells and presents a 'patchwork' appearance (Rossant and Frels 1980; Meinecke-Tillmann and Meinecke 1983; Gardner, R. L. 1968). These experiments demonstrated that sheep and goat blastomeres can form chimeric blastocysts and that such interspecies embryos are viable and may give rise to animals which are sheep-goat chimeras. The experiments also demonstrated that a goat foetus can develop to term in a sheep, and a sheep foetus can develop to term in a goat.

Successful development of interspecies chimeras through gestation and to adulthood has also been reported in the following species: house mouse-Ryukyu mouse (Rossant and Frels 1980), cow-zebu (Summers et al. 1983; Williams, T. J. et al. 1990), and sheep-cow (Fehilly and Willadsen 1986). Other experiments have shown that during early development of the sheep-goat blastocyst chimeras, increasing the proportion of transplanted cells in the inner cell mass can bias donor- or host-specific characteristics (Polzin et al. 1987).

Thus, if human cells greatly outnumbered host cells in an early blastocyst of a developmentally similar host (such as a chimpanzee), it is conceivable that substantial and ethically sensitive human-like characteristics might emerge in the resulting chimera (Karpowicz, Cohen and Van der Kooy, 2004). But to do these experiments, an embryo would have to be transferred into an uterus since there is, at present, no method for keeping human embryos alive *in vitro* beyond the 14-day stage.

Human-nonhuman chimeras created through xenotransplantation

Xenotransplantation (the transplantation of cells, tissues and organs from one species to another) was first considered almost 100 years ago. Since then, there have been sporadic instances of clinical applications but interest was only rekindled in the early 1990s as a result of new progress in the biomedical sciences. Indeed, because of the great success of human to human transplantation an ever increasing number of operations are being performed and the need for human cells, tissues and organs now greatly exceeds the supply. It is because of this shortage and the possibility for scientists to create a virtually unlimited supply of cells, tissues and organs through the use of nonhuman animal material, that xenotransplantation is currently being considered as a therapeutic solution to several previously incurable diseases relating to heart, liver, lung and kidney disorders. Additionally, there are other outstanding medical needs which could potentially be treated by xenotransplantation such as paraplegia due to spinal cord lesions and incurable neurological diseases including Parkinson's and Alzheimer's disease. Scientists in 1997 have already reported, in this regard, the transplantation of pig neural cells into a patient with Parkinson's disease (Deacon et al. 1997).

Xenotransplantation chimeras are also widely used in research and medicine when, for example, the transplantation of (1) human skin onto mice, (2) human tumours onto mice,[1] and (3) human bone marrow into mice is undertaken to provide appropriate models for biomedical examinations.

---

[1] For example, a common experiment to assess human embryonic stem cell quality and developmental potential is to try to use these cells to form teratomas on immunodeficient mice.

Specific ethical issues

Xenotransplantation raises medical, legal, cultural, religious and ethical issues. At first, public acceptance of such a procedure was minimal. With time, however, because of the increasing awareness by individuals for the desperate requirement of more organs and the potential for xenotransplantation to save lives, the procedure has become more ethically acceptable to most sections of society. This is provided that the medical problems of rejection and transmission of diseases have been addressed.

Human-nonhuman embryonic, foetal and post-natal chimeras

The usefulness and value of embryonic and foetal chimeras in research became clear about a decade ago in a series of remarkable experiments in which small brain sections from developing quails were removed and transplanted into the developing brains of chickens. The resulting chickens exhibited vocal trills and head bobs unique to quails, demonstrating that the transplanted parts of the brain contained the neural capacity for certain behavioural characteristics of quails. It also confirmed that complex behaviours could be transferred across certain species (Balaban 1997).

In this context, the discovery of human pluripotent stem cells, such as human embryonic stem cells in 1998, allowed researchers to consider related experiments that might reveal additional information about the manner in which embryos develop. This is because the cells found in five-day-old human embryos multiply prolifically and – unlike most adult cells – have the potential to turn into any of the about 200 cell types of the body.

As a result, the use of such pluripotent cells in embryonic research is extremely varied and promising. This is especially the case with respect to the creation of chimeras for research. In this regard, interspecies chimeras can be divided into 'primary chimeras' where the cells from the different species are combined into a blastocyst (50–100 cell early embryo) and 'secondary chimeras' where the cells have been combined at a later stage than the blastocyst. The different manners in which these chimeras are created are important since they may also, for instance, be associated with different ethical issues.

An example of primary embryonic chimeras in mouse models is the aggregation of different embryos before or at the eight-cell stage or by the injection of foreign embryonic stem cells into a mouse blastocyst. Which of the resulting tissues are of mixed genotype depends on the stage of the host embryo and the potency of the donor cells (Rossant and Spence 1998).

One particular kind of chimeric embryo in mice can be formed from the aggregation of early mouse embryos containing tetraploid cells (containing four sets of chromosomes which are generated by electrofusion of the cells of a normal two-cell embryo) and foreign mouse embryonic stem cells. Indeed, because the dysfunctional tetraploid cells are less able to develop into all the cells of the embryonic mouse (it is 'developmentally compromised'), the embryonic stem cells take over giving rise to live offspring which are completely formed from the descendants of the embryonic stem cells (Nagy et al. 1993; Wang and Jaenisch, 2004). While the practice of tetraploid complementation enables just a few injected embryonic stem cells to produce an entire foetus in mice it remains to be seen to what extent tetraploid cells might contribute to developing primates (Schramm and Paprocki 2004). Though very improbable, a question also remains whether the injection of human embryonic stem cells into similar tetraploid nonhuman embryos might lead to the dubious prospect of some interspecies chimeras yielding predominantly human foetuses.

On the other hand, some researchers have already injected human stem cells into normal nonhuman animal embryos before transferring these chimeric embryos into the wombs of animals. The proposed purpose of this research included the understanding of the mechanisms by which transplanted cells localize and differentiate in a host as well as using the stem cells in preclinical testing.

Another possible use of human-nonhuman chimeras is to test human pluripotency. To do this, human stem cells are injected into nonhuman blastocysts and allowed to develop beyond the 14-day stage in an animal's uterus or even to the stage of birth. It is likely, however, that some problems would arise with this approach if the animal species used was too distantly related and/or whose rate of development was much faster than that of the human species. A recent attempt to derive chimeras after the injection of human embryonic stem cells into mouse blastocysts was not very successful because the human cells were out-competed by the much faster dividing mouse cells (James et al. 2006). But such experiments might work if the host embryos originate from a closer matched animal species: a sheep or cow for developmental rate; or nonhuman primate (HCSTC 2007, HC272-II, Memorandum 61 Supplementary submission from Dr Robin Lovell Badge), though none of these human-nonhuman chimeric embryos could be successfully implanted into a human uterus without a human trophectoderm.[2]

In addition, it has been suggested that human-nonhuman chimeras could

---

[2] Trophectoderm: The cell layer from which the outermost layer of cells of the blastocyst differentiate and which attaches the fertilized ovum to the uterine wall and serves as a nutritive pathway for the embryo.

be developed with a 'humanized' immune system which could be used as models to study infectious diseases or for examining new vaccine candidates (Bhan, Singer and Daar 2010).

Human stem cells could also someday be grown into functioning tissue or organs in an animal for later transfer into a patient (Weiss 2004). Indeed, as there is a growing shortage of human tissue and organs to replace diseased and damaged ones, some researchers have already suggested creating certain kinds of animal chimeras on which human cells or organs are grown for use in transplantation.

For example, an international patent application has already been lodged, in 2000, for the production of human cells, tissues and organs using human-nonhuman chimeras (Townes and Ryan 2000). The procedure would involve knocking out in a nonhuman animal cell a gene required for the development of specific body parts of this animal.

This cell would then be introduced into an enucleated egg of a nonhuman animal from which the cell was obtained in order to create a cloned embryo which could not develop certain body parts. But when this embryo reaches the blastocyst stage of development, donor pluripotent human stem cells could be introduced and the blastocyst implanted into a pseudo-pregnant foster mother. The embryo would then develop into a chimeric animal that produced body parts from the human stem cells, but not the corresponding animal body parts. This is because the introduced human cells are out-competed in the development of the body parts of the animal, except for the body parts dependent for development on the gene (or genes) that were knocked out in the cloning cell used in the nuclear transfer.

In this way, an animal resulting from this blastocyst would contain the expected body parts of the animal from which the cloning cell was derived, except for the body parts dependent for development on the knocked out gene (or genes). Human cells, tissue(s) or organ(s) produced in the animals could then be taken from the animals and transplanted into patients in need of such cells, tissue(s),or organ(s).

Proof of principle of this approach was published in 2010 in Japan. Researchers injected rat stem cells into the early embryos of mice that had been genetically modified to not produce their own pancreas. When the mice were born, they were found to have developed a healthy pancreas that was almost completely composed of rat cells (Kobayashi et al. 2010).

Specific ethical issues

Although moral intuition about the creation of chimeras may vary, it is a subject of deep moral concern to many for whom the creation of animals with certain kinds of quantities of human cells, such as neural or germline cells, would be morally offensive. Accordingly, such research requires careful consideration and review (Karpowicz, Cohen and Van der Kooy 2004; Loike and Tendler 2008). This is especially the case when human pluripotent stem cells are used in the animal brain, whose biochemistry or architecture might be affected by the presence of human cells (Karpowicz, Cohen and Van der Kooy 2005). Similarly, care must be taken lest human pluripotent stem cells alter the animal's germline. The putative ethical significance of neural and germline tissue over and against other bodily tissue is something that requires further exploration, which will be conducted in the third part of the book. At this point the issue of concern is the more technical question of whether such experiments are feasible and whether it is likely that they would result in human characteristics being displayed and/or inherited.

Few scientists are eager to do these kinds of experiments without guidance because of the associated risks. For example, some human cells may find their way to the developing testes or ovaries, where they might grow into human sperm and eggs. If two such chimeric mice were to mate, a human embryo might form which would be trapped inside a mouse.

Accordingly, various precautions seem reasonable in studies that involve the transfer of human pluripotent stem cells into nonhuman animals and should be considered in any prior review of a protocol. Questions that have already been raised by the US National Academy of Sciences in its 2005 report entitled 'Guidelines for Human Embryonic Stem Cell Research', which was updated in 2007 and 2008, in this context, include (at p. 41):

Are human pluripotent stem cells required, or can cells, from other nonhuman species be used?

Has sufficient animal work preceded the proposed work involving human pluripotent stem cells?

If human pluripotent stem cells are to be transferred into an animal embryo or foetus, have studies (for example, with embryonic stem cells from other species or interspecies chimeras) suggested that the resulting creature would exhibit characteristics that would be ethically unacceptable to find in an animal?

If visible human-like characteristics might arise, have all those involved in these experiments, including animal care staff, been informed and educated about this?

In the light of these questions, the National Academy of Sciences in the USA in its updated report of May 2010 entitled 'Guidelines for Human Embryonic Stem Cell Research', recommended (in Sections 1.3a –1.3c, 6.7, 7.3 and 7.5), that:

Research involving the introduction of human embryonic pluripotent stem cells into animals other than humans or primates at any stage of embryonic, foetal, or postnatal development should only be permissible after authorisation from an appropriate and independent ethics committee.

Research involving the introduction of human embryonic pluripotent stem cells into nonhuman primates at any stage of foetal or postnatal development should only be permissible after authorisation from an appropriate and independent ethics committee.

Research in which there is a significant possibility that the implanted human (non-embryonic) pluripotent stem cells could give rise to neural or gametic cells and tissues in the nonhuman animals, at any stage of development or maturity, should only be permissible after authorisation from an appropriate and independent ethics committee.

In the above three cases, particular attention should be paid to the:

(1) Extent to which the implanted cells colonize and integrate into the animal tissue,
(2) Degree of differentiation of the implanted cells, and
(3) Possible effects of the implanted cells on the function of the animal tissue (consideration of any major functional contributions to the brain should be a main focus of review).

Introduction of human embryonic pluripotent stem cells into nonhuman mammalian blastocysts should be considered only under circumstances in which no other experiment can provide the information required.

Research in which human embryonic pluripotent stem cells are introduced into nonhuman primate blastocysts or in which any embryonic nonhuman stem cells are introduced into human blastocysts should not be permitted at this time.

No animal into which human embryonic pluripotent stem cells have been introduced such that they could contribute to the germ line should be allowed to breed.

Moreover, the International Society for Stem Cell Research (ISSCR), an independent, non-profit organisation involving groups from 29 countries highlighted in its 2006 'Guidelines for the conduct of human embryonic stem cell research' a number of points of concern regarding chimeric animals containing human cells (ISSCR 2006: 7) including:

The degree of the resulting chimerism;

The type of tissues that are chimerised;

The fact that the earlier human stem cells are introduced during animal development, the greater the potential for their widespread integration;

The fact that the introduction of a greater number of cells later in development may have an equivalent effect to the above;

Whether implanted cells might migrate through the animal's body.

Because of this, the Guidelines state that when considering applications for this type of research, the responsible regulatory body should pay special attention to:

(a) The probable pattern and effects of differentiation and integration of the human cells into the non-human animal tissues; and

(b) The species of the animal, with particular scrutiny given to experiments involving nonhuman primates. Experiments that generate chimerism of the cerebral cortex or germline should be subjected to especially careful review (AMS 2007: 34–5).

The ISSCR guidelines also indicate (at pp. 6–7) that:

Research that generates chimeric animals using human cells, including (but not limited to) introducing totipotent or pluripotent human stem cells into nonhuman animals at any stage of post-fertilisation, foetal or post-natal development is permissible only after additional and comprehensive review by a specialised body. Indeed, such forms of research require the provision of greater levels of scientific justification, consideration of social and ethical aspects of the research and reasons for not pursuing alternative methods.

Implanting any products of research involving human totipotent or pluripotent cells into a human or nonhuman primate uterus should not be permissible at the current time.

Research in which animal chimeras, incorporating human cells with the potential to form gametes, are bred together should not be permissible at the current time.

Incorporation of human stem cells into post-natal nonhuman animals

Some scientists have suggested that the implantation of human stem cells (including pluripotent stem cells) into experimental animals would provide useful experimental results. These could include experiments for assessing the contribution of stem cells to repair a damaged spinal cord and determining the contribution of these cells to integrate into damaged muscular tissue of the heart (HCSTC 2007: 50). However, even if this is regarded as acceptable, a much more problematic question is whether human stem cells could also be implanted into more ethically sensitive parts of the nonhuman animals, such as the brain or those parts responsible for the development of gametes.

Experiments already undertaken

Genetic human-mouse chimeric mice

In 2008 a team of scientists from Brazil extracted adult stem cells from the soft material in the centre of the teeth of a human male donor. These were then injected into the testes of live male mice for further development. The mice were subsequently killed at various intervals and their testes examined. The researchers found that the human stem cells had allegedly not only settled into the testes of the mice but had also successfully differentiated into cells that resembled viable human sperm (Leake and Templeton 2008).

Genetic human-monkey chimeric monkeys

In 2008, researchers in USA transplanted human neural progenitor cells into the brains of nonhuman primates to study Parkinson's disease (Emborg et al. 2008). Other teams have since continued this line of research with a relative amount of success (Redmond 2010).

Specific ethical issues

The ethical concerns commonly raised in the incorporation of human stem cells into post-natal animals relate to the possibility that such cells, because of their potency, could give rise to cells of the germline or the brain.

In response to this, it should be noted that it is highly unlikely that human stem cells could contribute, in an unplanned manner, to the germline after implantation into a post-natal animal because the germline is set aside very early in foetal development (National Academy of Sciences 2005: 33).

On the other hand, the contribution of human stem cells to the brain of a nonhuman animal is more difficult to evaluate. One purpose for introducing human stem cells or human neural progenitor cells into the brain of a nonhuman animal is to provide a model for the repair or regenerative processes and to yield neurons which may, for example, contribute to combating neurodegenerative diseases.

The idea that human neuronal cells might participate in 'higher-order' brain functions in nonhuman animals, however unlikely that may be, raises concerns that need to be considered (National Academy of Sciences 2005: 33).

It is indeed particularly difficult to predict the consequence of transplanting human brain stem cells into nonhuman primates. If such a chimera was created, then it is not improbable that the developing brain would increasingly exhibit functions particular to human behaviour. The distinction between humans or nonhuman nature would then be even more difficult to evaluate and the moral status of the being would become uncertain (Deutscher Ethikrat 2011: 112). Because of this, such experiments should only be undertaken after careful evaluation by an interdisciplinary ethics committee.

Incorporation of (1) Human Stem Cells into Post-blastocyst Stages of Nonhuman Embryos or (2) Nonhuman Stem Cells into Post-blastocyst Stages of Human Embryos

Experiments incorporating human stem cells into appropriately organized nonhuman embryos or foetuses would offer greater opportunities to reveal the potential of such cells, which would be marked by their genetic composition. Such experiments have already been undertaken in testing the capacity of neuronal progenitor cells derived *in vitro* from mouse embryonic stem cells by transplanting them into chicken embryos.

In the UK, the Human Fertilisation and Embryology Act 1990 did not have any provisions which could specifically address the creation of human-nonhuman chimeric embryos. However, the Human Fertilisation and Embryology Act 2008 which amended the 1990 Act sought to address this legal vacuum by indicating that no person shall bring about the creation of a human admixed embryo except in pursuance of a licence. In this regard, a human admixed embryo includes (1) any human embryo that has been altered by the introduction of one or more animal cells or (2) any embryo which contains both (nuclear or mitochondrial) DNA of a human and (nuclear or mitochondrial) DNA of an animal but in which the animal DNA is not predominant. The legal meaning of 'predominant' has already been discussed above, in relation to national legislation. The ethical question of whether it is significant if human DNA predominates is an issue that requires further exploration and is related to the issue of whether certain tissues (neural and germline) are more ethically significant than others.

Experiments already undertaken

Genetic human-mouse chimeric foetuses

Scientists at Stanford University have injected human neuronal stem cells into mouse foetuses, creating mice whose brains were about 1 per cent human. By dissecting the mice at various stages, the researchers were able to see how the added brain cells moved about as they multiplied and made connections with mouse cells (Weiss 2004). The same scientists now want to add the dysfunctional human brain stem cells that cause Parkinson's disease, Lou Gehrig's disease and other brain ailments and study how those cells make connections. Indeed, scientists suspect that these diseases, though they manifest themselves in adulthood, begin when something goes wrong in early development.

The Stanford team also considered the creation of chimeric mice whose brains are 100 per cent human. They suggest, however, that if the brains seemed to be taking on a distinctly human architecture – a development that could suggest a specific amount of 'humanness' – then the mice could be killed. On the other hand, if the brains looked as if they were organizing themselves in a specifically mouse brain architecture, then the mice could be used for research (Weiss 2004). In this case there may be an ethical discrepancy, in that killing is considered more justifiable when increasingly human qualities are observed, rather than according an increasing right to life with increasing signs of humanity.

Another team of scientists, in the USA and Japan, reported a similar experiment, in October 2005, whereby human embryonic stem cells were implanted into the brains of foetal mice 14 days after conception (gestation of mice usually lasts about 21 days). These were then shown to develop into active human neurons and integrate into different parts of the brains of live adult mice (Muotri et al. 2005).

The same year, in January 2005, an informal ethics committee at Stanford University had endorsed the proposal to create mice with brains made nearly completely of human brain cells. The chairperson of this committee indicated that the board was satisfied that the size and shape of the mouse brain would prevent the human cells from creating any traits of humanity. But just in case, the committee recommended closely monitoring the mice's behaviour and immediately killing any that displayed human-like behaviour (Associated Press, 2005).

Genetic sheep-human chimeric foetuses

In 2001, researchers at the University of Nevada, USA, injected human stem cells coming from bone marrow or umbilical cords into sheep foetuses. The sheep then grew up with a small proportion of human cells throughout their bodies (Goodman 2001).

A few years later, in December 2003, it was announced that human stem cells which were injected into sheep foetuses were able to produce a surprisingly high proportion of human cells in some organs. In most cases between 7–15 per cent of all the cells in the sheep's liver were human though a few had as much as 40 per cent of human cells (Associated Press 2005). The human cells were injected around halfway through gestation – before the foetus' immune system had learned to differentiate between its own and foreign cells, so that the animal does not reject them, but after the body had formed. This procedure ensures that the resulting animals look like normal sheep rather than sheep-human combinations. The researchers recognized, however, that there was no way for them to determine whether the sheep foetuses had human brain cells (New Scientist 2003).

Genetic monkey-human chimeric foetuses

In 2001, researchers indicated that they had implanted human neuronal stem cells into the skulls of unborn monkeys. They then showed that these cells were incorporated into the developing brains of the animals (Associated Press 2001; Václav et al. 2001; Redmond 2002).

Genetic pig-human chimeric foetuses

In January 2004, pigs grown from foetuses into which human stem cells were injected were shown to be made up of three kinds of cells. In other words, researchers indicated that they were made up of (1) pig cells, (2) human cells and (3) hybrid cells, the latter being fully fused pig-human cells in which the DNA from both species were mixed at the most intimate level (New Scientist 2004).

Genetic rat-human chimeric embryos

In 2005, researchers were able to show that human adult stem cells from bone marrow, when placed in a rat embryo, integrated into the developing rat kidney. The integrated cells were shown to have differentiated into complex functional kidney structures (Yokoo et al. 2005).

Genetic goat-human chimeric foetuses

A team of Shanghai scientists reported, in May 2006, that it had performed the world's first successful transfer of human stem cells into 45–55 days' old goat embryos. The scientists transplanted stem cells taken from human umbilical cords into embryos carried by 50 goats. When the offspring were born, 39 were found to have some human genetic characteristics in their blood and internal organs (Huang et al. 2006).

Genetic chicken-human chimeric embryos

Israeli Scientists demonstrated, in 2002, that human pluripotent stem cells could be grafted into 1.5 to 2-day-old chick embryos (the gestation time for chickens is about 21 days). The grafted human cells survived, divided, differentiated and integrated in the chick hosts which were killed before hatching (Goldstein et al. 2002).

Specific ethical issues

Again, ethical sensitivities arise concerning neuronal and germline cells and are perhaps even more of a concern than in the case of transplantation into post-natal animals. This is because the human stem cells might be expected to have a greater opportunity to participate in the development of the biological entity. Thus inasmuch as the 'precautionary principle' applies to this area of research, it will do so more strongly in the case of pre-natal than post-natal interventions.

Incorporation of (1) Human Pluripotent Stem Cells into a Nonhuman Blastocyst or its Preliminary Embryonic Stages or (2) Nonhuman Pluripotent Stem Cells into Human Blastocysts or its Preliminary Embryonic Stages

A nonhuman animal blastocyst or its preliminary embryonic stages into which human pluripotent stem cells are transplanted raises difficult ethical issues because the developing inner cell mass, which may eventually become the foetus, could consist of a combination of human and nonhuman cells.

At present, it is not possible to predict the extent of human contribution to such chimeras. If the recipient embryo was from a nonhuman animal that is biologically close to a human, the potential for human contributions may be greater (National Academy of Sciences 2005: 34) though preliminary research suggests that the creation of primate chimeras using such a procedure is unlikely (Tachibana 2012). This is in contrast, however, with the possible successful combination of very early human and nonhuman primate embryos.

For example, if such chimeras were made between human beings and nonhuman primates then the likelihood that these would give viable embryos would be considerable. However, it would not be possible to indicate whether the different cells were competent to differentiate very far if they were destroyed before the 14-day limit after the creation of the embryo. But by taking individual cells (or parts) of these destroyed early embryos and culturing them or using them to create further chimeras, some useful information may be obtained.

Some scientists have also suggested that there may be occasions when it would be useful to allow the development of certain types of human-nonhuman chimeras beyond 14 days in an animal's uterus, and even for the chimeras to be born. The proposed experiments would involve the introduction of 'marked' human stem cells into the developing blastocyst stage embryos of a nonhuman animal. These embryos could then be implanted into a surrogate uterus to further determine pluripotency by looking for the presence of marked cells throughout the organs and tissues of the developing embryo, foetus or organism which has been born (HCSTC 2007: 51).

This possibility was supported by the UK House of Commons Science and Technology Committee in its 2007 report entitled 'Government Proposals for the Regulation of Hybrid and Chimera Embryos' in which the Members of Parliament indicated in a paragraph entitled 'Development of human-animal chimera or hybrid embryos past the 14-day limit in vivo' that 'legislation [should] allow for regulation of the implantation of human stem cells, whether created from human embryos or human-animal chimera or hybrid embryos, into animal blastocysts' (p. 51).

In the UK, the Human Fertilisation and Embryology Act 1990 did not have any provisions which could specifically address the creation of human-nonhuman chimeric embryos. However, the Human Fertilisation and Embryology Act 2008 which amended the 1990 Act sought to address this legal vacuum by indicating that no person shall bring about the creation of a human admixed embryo except in pursuance of a licence. In this regard, a human admixed embryo includes (1) any human embryo that has been altered by the introduction of one or more animal cells or (2) any embryo which contains both nuclear or mitochondrial DNA of a human and nuclear or mitochondrial DNA of an animal but in which the animal DNA is not predominant. The creation of human-nonhuman chimeric embryos in which nonhuman DNA is predominant generally remains in a legal vacuum.

Experiments already undertaken

Genetic human-mouse chimeric embryos

In 2003, scientists at the South Korean firm Maria Biotech, were reported to have allegedly injected human embryonic stem cells labelled with a fluorescent protein into 11 mouse blastocysts which later developed. The embryos were then carried by foster mice, whereby five offspring were born with fluorescence in tissues including the heart, bones, kidney and liver. However, the scientists terminated the project after having to address 'severe protests' from the public (Boyce 2003).

Other researchers from the USA showed, in 2006, that human embryonic stem cells could engraft into mouse blastocysts, where they proliferate and differentiate *in vitro* and persist in mouse-human embryonic chimeras that implant and develop in the uterus of foster mice. Interestingly, the scientists found that the human embryonic stem cells engrafted seemed to prefer occupying the inner cell mass of the mouse blastocysts which parallels that of their origin in human embryos. They did not contribute to the cells of the outer layers of the embryo. However, the researchers did indicate that the influence of these human embryonic stem cells seemed to disrupt, in most cases, the development of the human-mouse embryonic chimeras (James et al. 2006).

In a way, these results were predictable since human embryonic cells divide at a lot slower rate than those of mice and will be rapidly overwhelmed by the latter.

Specific ethical issues

It is very probable that the creation of early human-nonhuman embryonic chimeras would create grave and complex ethical difficulties. This is reflected in the UK Animal Procedures Committee Report on Biotechnology published in 2001 which indicated that 'No licences should be issued for the production of embryo aggregation chimeras especially not cross-species chimeras between humans and other animals' (Animal Procedures Committee 2001, Recommendation 5). Furthermore, these ethical difficulties would seem to be exacerbated depending on the tissues transplanted or generated and the nonhuman animal used. As the Academy of Medical Sciences indicated in its 2011 report entitled 'Animals Containing Human Material', in the context of creating embryonic and foetal chimeras, ethical concerns 'may increase significantly' if nonhuman primates are used (AMS 2011: 40).

Thus, the precautionary principle would seem to imply that at least some experiments in creating early human-nonhuman embryonic chimeras should be considered as deserving to be proscribed. Nevertheless, specific conclusions will require a more thorough ethical analysis and this in turn will depend on a deeper reflection on the concepts, distinctions and principles involved. Such a task is not only technical but requires engagement with cultural, worldview and ethical perspectives.

PART THREE

Cultural, worldview and ethical perspectives

8

Cultural perspectives

Assessing cultural attitudes to human-nonhuman combinations is a complex task, which of course is made all the more complicated by the difficulty of defining the nature and content of the concept of culture as well as the cultural attitudes to human-nonhuman combinations.

The term of 'culture' usually refers to a multitude of aspects in a given society including the mutual, intellectual, ethnic, moral, artistic and spiritual characteristics. Almost every person is part of a number of different, often competing cultures, depending on their context, background, family, education, experiences and work.

Public attitudes to new technology

Public opinion is often based on the cultural values of large sections of the general public. And in this regard, it should be noted that the actual content of cultural values in the biomedical field reflects a whole spectrum of attitudes towards science, technology and medicine. These attitudes are probably also influenced by historical factors such as the Judeo-Christian heritage in many European countries.

While many people are ready enough to accept the benefits of modern science and medicine, and there is generally no generic public distrust of science, there is nonetheless a growing concern relating to the risks and dangers accompanying specific biomedical discoveries. The widely publicized cases of Bovine Spongiform Encephalopathy (BSE), Creutzfeldt-Jakob Disease (CJD) and the ongoing questions about food safety and genetically modified (GM) food, have created an atmosphere of distrust of science and scientists in some quarters of society. There is a fear that no one is really 'in control' or 'knows what will happen' and doubts remain concerning the suggestion that governments can actually prevent or control potential lasting negative consequences.

In addition, there is often a negative emotional reaction amongst the general public towards certain technologies. This is frequently referred to as the 'yuck' factor, which should not simply be dismissed in ethical considerations as irrational or sentimental concerns. Rather, it should be acknowledged that such reactions may well reflect an underlying but inarticulate social intuition. If people feel an emotional revulsion towards a procedure this may be important and relevant to the moral positions, deeply held beliefs and intuitions of a society.

As Karpowicz et al. indicate 'it is important to acknowledge that taboos based in <u>repugnance</u> and intuition play a significant role in preserving core social values within most societies' (Karpowicz, Cohen and Van der Kooy, 2005: 111). Here, repugnance is defined as a reaction that may give rise to major societal disturbance and disorder in contrast to a sense of repulsion which does not have such drastic consequences. Kapowicz et al. also recognize that these taboos may be considered as social conventions that are the result of different historical and cultural contexts and, as such, may be subject to changes as the context in which they arise alters.

The common and intense revulsion against creating human-nonhuman combinations does not settle this issue, but it does provide a starting point. Leon Kass, who was a past chairperson of the US President's Council on Bioethics, in a well-known essay 'The Wisdom of Repugnance', states in 1998 that, 'revulsion is not an argument; and some of yesterday's repugnances are today calmly accepted ... [i]n crucial cases, however, repugnance is the emotional expression of a deep wisdom, beyond reason's power to fully articulate' (Kass 1998: 18).

In a cogent examination of the same subject, the philosopher Mary Midgely warns against thinking of feelings either as though they had no rational object, or of reason as though it was, or should be, unaccompanied by feelings. If persons seriously judge something to be wrong then strong feelings will accompany that judgement. Emotional reactions may be appropriate, or they may be inappropriate, and to decide which '[w]e must spell out the message of the emotions and see what they are trying to tell us' (Midgely 2000: 9).

However, a spontaneous 'yuck' reaction can also be exploited for many different purposes, including its 'entertainment value'. This happened in modern science fiction treatments, such as H. G. Wells' *Island of Dr. Moreau* which portrayed a fictional island where exiled scientists combined humans and beasts through vivisection. Films relating to cannibalism have also been made which use the perceived revulsion of the general public towards persons eating their own kind (i.e. other human beings as opposed to other species).

Of course, novels and films do not always represent scientific reality, and many scientists now regret the influence of Mary Shelley's classic novel

Frankenstein. However, it may be that the Frankenstein story still has a place in the context of a human-nonhuman combination debate in that it seeks to explore, express and represent some of the revulsion, anxieties and emotions related to crossing biological boundaries.

More recently, black horror films such as *The Fly* and *Splice* (Bradshaw 2010) have also portrayed the frightening prospects of what can go wrong when scientists, working in secret and without any ethical oversight, end up creating new beings which can only be considered as human-nonhuman 'monsters'.

The basis of the revulsion and taboos relating to the creation of human-nonhuman combinations are probably the result of the view that biological elements that are different should be kept apart since mixed entities do not fit neatly into existing categories. From this perspective, animal combinations, such as a pig with a chicken's head, which cannot be clearly put in a specific box are usually viewed as monstrous, not merely because of their hideousness (which is merely an aesthetic appearance of a lack of wholeness), but because they are not integrated wholes.

Interestingly, it may be the external appearance of the interspecies entity which would create the most aversion amongst the general public in contrast to any mixing of non-visible internal organs (AMS 2011: 79). This is because the public would be instantly confronted with its inability to identify the important visible distinctions between species which has crucial consequences on the manner in which a living being is classified (AMS 2011: 55–6). The entity would be a something-in-between and may be deemed to have no place in society. Such feelings are obviously heightened when one of the parts is human, since additional questions of identity, legal rights and psychology come into play.

In this context, even amongst the enlightenment philosophers, John Locke (1632–1704) struggled to define the moral nature of 'half beast and half man' monsters which he believed were the result of unnatural sexual practices (Locke III.VI.23 and IV.IV.16). In his reflection on this problem, he could not bring himself to conclude that the outward shape of the being necessarily implied that it did not have an 'internal constitution' like that of any other human person (Locke III.VI.22). To the question of the exact nature of the 'changelings' 'between man and beast', he acknowledges that their existence may be used by some to undermine religion but in the end could not give a final decision, indicating that 'To their own master [a faithful Creator and a bountiful Father] they stand or fall' (Locke IV.IV.14).

A reflection on the repugnance towards such beings as expressed by the 'yuck' factor can also be recognized in the legislation of a number of countries. For example, the UK Human Fertilisation and Embryology Act 1990 indicated that completely human embryos created for research had to be destroyed within 14 days after coming into existence. But hybrid embryos created from the mixing of animal eggs with human sperm had to be destroyed no later than the two cell stage (1–2 days after their creation), without any rational reason being given for the disparity. The Australian Prohibition of Human Cloning and Regulation of Human Embryo Research Amendment Act 2006, which was enacted later, makes a similar distinction.

In order to understand public repugnance it is necessary to explore a fundamental difference in philosophical worldviews. According to the materialist and molecular reductionist worldview held by many (though not all) scientists, biological beings are just made up of several types of complex macro-molecules, which are composed of strings of small molecules which are common to all species. The only differences between species are merely the result of minor differences in the ordering of small molecules within the long chains. For instance, the difference between a protein from cattle and human beings could be completely described by compiling a catalogue of the genetic differences that code for the proteins.

This worldview does not recognise the idea of qualitative boundaries in nature: nature looks rather like a well-blended soup. Within this paradigm, species differences are a matter of drawing an arbitrary line, and are to some degree illusory and unreal, a matter of quantitative not qualitative differences.

Such a perspective is clearly apparent in the statements that stem cells derived from certain kinds of hybrid embryos are 99.9 per cent human and 0.1 per cent animal (Sample 2006) or even that, rounding up, they are simply 'human'. Like genetic engineering, the mixing of human DNA and animal eggs is less problematic from this standpoint, because the two are not 'really' different in kind. Yet this 'reductionist' worldview is itself open to criticism as it leaves no room for the human meaning of this specific form of life. The bioethicist Alastair Campbell has recently argued that 'it is morally impoverishing to think about these [bioethical] issues without a full account of ourselves as embodied selves' (Campbell 2009: 25).

From a more historical perspective, moral emotional revulsion has also taken place with the creation of a number of transgenic animals that suffered unexpected side effects such as those experienced by the 'Beltsville pigs' (Mench 1999). Many persons are likely to be concerned about the animal welfare implications of experiments like this one. The Beltsville pigs and other experiments which tend in a similar direction have given rise to a strong interest in, and concern about, animal rights and welfare. Groups concerned with the protection of animal interests and the successful use of publicity by some vocal activists have gathered media attention and influenced public opinion.

More recent developments, such as genetically modified micro-organisms for treating toxic wastes or coping with extreme environments, have also met with opposition. This has not originated so much from the general public but

from environmental groups concerned about the inability to control genetically engineered organisms once released into the natural environment. The
environmental and green lobbies have been particularly effective in building
up and mobilizing the public's disquiet and emotional revulsion, basing their
opposition on the precautionary principle – that one should not introduce
new technologies or applied science until and unless one can be sure of the
consequences.

As far as limits to scientific manipulations of nature are concerned, an
important distinction should be made between different types of interspecies
combinations. Many scientists have assumed that it is often enough to
reassure the public that no 'monster' will be born and that all they are doing
is merely manipulating a few cells. But paradoxically, public concern about the
proposed experiments is, in all probability, to some extent heightened rather
than diminished by the fact that in the case of creating human-nonhuman
embryonic combinations, one is addressing the manipulation of embryos
rather than nonhuman animals or human beings. If nonhuman tissue is transplanted into an adult human being, the result will unquestionably be a human
being. Manipulations at the level of organs or fully developed organisms
(other than brain or reproductive organs) do not threaten the identity or
species status of the resulting entity. A human being who has had nonhuman
tissue transplanted will undoubtedly remain human.

Manipulations at the subcellular level, however, which can only be achieved
by intervening at the earliest stage of an organism's development, have the
potential to cause far more profound changes in the resultant physiology
and behaviour of the fully developed organism. This then raises questions
about the species status of the resulting entity. It is the vulnerability of the
embryo to such profound manipulation which attracts concern, and raises
valid questions about whether there should be limits imposed on scientists'
manipulation of nature.

There is also a concern that financial motives may be a key driving force.
Indeed, there is a fear that profit and economic factors may be driving the
move to human-nonhuman combinations rather than the genuine benefit
or needs of society. In other words, that economic motives may end up
superseding ethical concerns. In one of the recommendations from the
2007 UK House of Commons Science and Technology Committee entitled
'Government Proposals for the Regulation of Hybrid and Chimera Embryos' it
was stated that a 'ban and the prospect of a ban in draft legislation on human-
animal chimera or hybrid embryos would undermine the UK's leading position
in stem cell research' (HCSTC 2007: 54). In the same vein, those in favour
of permitting human-nonhuman combinations expressed concern that some
biotechnological or pharmaceutical companies may move their research and
experimentation bases to countries where regulations are more permissive.

This was also reflected in one of the recommendations from the 2007 UK House of Commons Science and Technology Committee entitled 'Government Proposals for the Regulation of Hybrid and Chimera Embryos 'which indicated that 'a ban or a proposed ban may not only encourage researchers to leave the UK in order to undertake their research in a more permissive regulatory regime, but it may also inhibit early stage researchers entering the field' (HCSTC 2007: 55).

This economic motivation could have a detrimental effect on public trust in pronouncements by scientists. In a survey of Public Attitudes to Science in 2008 commissioned by the United Kingdom Research Councils, 78 per cent of people thought that the independence of scientists is often put at risk by the interests of their funders (People Science & Policy 2008, Table 3.16).

Necessity for public debate

Though books and films, such as *Frankenstein* and *The Island of Doctor Moreau* may be recognized by the general public, it remains unclear whether these imaginary explorations have been influential on their ethical reflections concerning human-nonhuman interspecies entities.

During the debate surrounding the passing of the Human Fertilisation and Embryology Act 2008 there was considerable frustration among parliamentarians that there was little good evidence for the actual state of public opinion about the creation of human-nonhuman embryos. The UK House of Commons Science and Technology Committee complained that:

> We have seen no conclusive evidence to indicate the true state of public opinion on the creation of animal-human chimera and hybrid embryos for research purposes ... **We find it unhelpful that witnesses on both sides of the argument have claimed to represent the public view, where supporting evidence for this is lacking.** (HCSTC 2007, para 113, emphasis in the original).

Similarly the UK House of Parliament Joint Committee preparing the Bill in 2007 was 'concerned by the unsubstantiated claims made about public opinion and public support and by the lack of evidence provided' (Joint Committee 2007, para. 22).

It was to remedy this lack of evidence that the Human Fertilisation and Embryology Authority (HFEA) in the UK undertook a three month consultation exercise in 2007 on the topic of human-nonhuman hybrids and chimeras. It realized that the general public was only beginning to reflect on these issues. In addition, comments were received from the public about the lack of information and even the perceived misinformation being presented by medical researchers including implicit or explicit claims about the supposed benefits of such combinations (HFEA 2007b, paragraph 5.40).

The HFEA report was too late to inform the Joint Committee which drew up the Bill, but it provided valuable evidence which is still relevant to the question of human-nonhuman entities. This showed that there was a significant initial reaction against the proposal, but that when the prospective benefits of the research were shared with the public then a considerable section of the population was prepared to revise their views. This was especially the case for those forming their opinion on the issue for the first time. This meant that whilst some members of the public initially reacted with disgust, their opinion often shifted significantly after hearing information provided by those who wished to do the research and having a chance to discuss the issues with them (HFEA 2007b, paragraph 6.6).

The HFEA's study remains the most detailed investigation of public opinion in the UK on human-nonhuman combinations. But the HFEA has been heavily criticized for the manner in which the consultation was performed. The bioethicist Françoise Baylis has presented credible evidence that a view had already been taken by the Authority and the consultation was not so much a meaningful engagement with the public but, instead, an exercise in public relations. She also emphasized that although the 'HFEA's rule-guided and strategic modes of public consultation on the ethical and social implications of creating human/animal embryos in research may be legitimate in a strict sense, they fall far short of embracing the democratic ideal of input-oriented legitimacy' (Baylis 2009). Serious concerns were, therefore, raised about this public consultation and the manner in which the HFEA sought to assess the views of the people it approached. In addition, it is generally recognized in public opinion polling that the framing of a question affects the manner in which people reply (Bonnicksen 2009: 8).

Moreover, the HFEA did not investigate whether the acceptance of human-nonhuman research would be maintained in circumstances where the benefits were to appear less certain or if alternative avenues of research seemed to offer similar or better prospects. This is unfortunate for it is not an idle question. Nevertheless, the HFEA admitted that public opinion was opposed to the creation of human-nonhuman hybrid embryos unless these were likely to lead to medical advances. There was no 'blank cheque' in terms of unconditional support for this research.

Part of the pressure that skewed the HFEA consultation on hybrids was the intense media campaign at the time in favour of legalizing cytoplasmic hybrids (see also the discussion above in relation to UK national legislation). The aim of the science lobbyists was to overcome public misgivings (Jones 2009a) and enable research on the entire range of human-nonhuman combinations. Their strategy relied first on the explicit claim that cytoplasmic hybrids were 'a *vital* tool to advance the progress of research into the potential of embryonic stem cells' (AMRC [Association of Medical Research Charities] 2008, emphasis added). Secondly it rested on explicit or implicit promises that, if cytoplasmic hybrids could be used, scientists would be 'close to the breakthroughs that … can save and improve the lives of thousands and, over time, millions of people' (Brown 2008). Thirdly it relied on the deliberate conflation of every kind of human-nonhuman combination with cytoplasmic hybrids. There was virtually no public debate on true hybrids or chimeras.

Other bodies were also involved in the campaign. In addition to the Science Media Centre and the the UK Association of Medical Research Charities, the Medical Research Council, which distributes tax-funding to medical research in the United Kingdom also argued for the vital importance of this research. However in January 2009 the same Medical Research Council turned down funding to the only two research groups who had applied to work on admixed embryos. At least one significant factor behind this shift was the need to fund new developments in induced pluripotent stem cells. This exciting breakthrough had a marked impact on the whole research environment. However, the relative merits of these avenues of research were not well understood by the public. It seems that the effort to persuade the public to back permissive legislation created a misperception of the scientific merits of human-nonhuman embryo research.

The public and parliamentarians were thus very poorly served by the media debates on the content of the 2008 Act (Jones 2009b). There was very little reflective discussion of the perplexing nature of human-nonhuman combinations. The result was an Act which allowed the whole range of human-nonhuman admixed organisms if they were restricted to the embryonic stage. Nevertheless, while in one way the legislation was very permissive, there was at least some concern that all these admixtures should fall within the legislation and should require a license from the HFEA.

In parallel to the gathering of evidence of public opinion in the UK, there have also been attempts to gather similar data internationally. Thus the Second BBVA Foundation International Study on Biotechnology, published in May 2008 (BBVA 2008: 28) added a question on hybrid embryos to its survey of international opinion. The results of this were consistent with the HFEA study, with the initial reaction of most people, in a majority of countries, being one of disapproval. In fact, the United States, Germany and Japan all showed higher levels of disapproval than Britain.

In summary, there is good evidence that the initial reaction of most people in developed countries is to oppose the creation of mixed human-nonhuman embryos. The HFEA data suggests that this reaction is overcome only inasmuch as the research is believed to be necessary for medical advances in relation to named diseases (Jones 2009a). At the same time as the science advances, the case for the 'necessity' of human-nonhuman embryo research seems to be getting not stronger but weaker.

This underlines the need for all involved in the human-nonhuman combination debate, including the general public, to understand the arguments both for and against these types of experiments in order to make an ethical decision. Generalized assumptions may be misleading. For example, in a report entitled 'Hype, hope and hybrids' published in June 2009 by the Science Media Centre, it was revealed that ethical arguments in favour of using human-animal embryonic combinations were not very present in the press. It added, in a very pragmatic way, that 'This may reflect an unspoken assumption that research devoted to the relief of human suffering is of its nature an ethically desirable enterprise' (Watts 2009: 42).

It is important that the public should be given the necessary and appropriate information in order to coordinate a genuine debate about current attitudes towards human-nonhuman combinations, animal issues and human concerns. Openness in the activities of the regulatory bodies, advisory groups and working parties especially in relation to the scientific issues of risks, benefits, safety and monitoring of humans and nonhuman animals is also vital if a proper informed debate is to take place. In other words, if this type of technology is to be accepted and used, a number of questions must first be answered. These include demonstrating that:

There is a genuine biomedical need for human-nonhuman combinations;

There are no appropriate alternative options which are available;

The technology is effective and reliable;

The highest possible levels of safety for patients and the wider human population can be achieved;

All issues of animal husbandry, care, welfare and use are strictly monitored;

There are real benefits for patients, families and society and not just for the commercial companies who stand to make a profit from this procedure;

There are no other serious ethical reasons that would give rise to concerns; and

The procedure would not result in considerable offence being taken by a significant proportion of the general public.

It is therefore critical that the public have a good understanding of the issues involved and that the various media play their part in creating a genuine dialogue. A good deal of thought needs to be put into how public debate will be carried out as well as by whom and on what basis it will be delivered. If the scientific community acts with disregard to the general public and if concerns are dismissed as irrational mystifications or sentimentalism, then they will erode political public support. This is especially the case if sweeping claims are made for benefits which then fail to materialize.

Thus, it is necessary to sympathetically examine and seek to understand and address (rather than dismiss without serious reflection) the sources of these anxieties. In the past, the usual division of morality between individual autonomy and quantifiable public utility has failed properly to predict and motivate a response to predicaments, such as the environmental crisis. It has also been suggested that the manner in which ethical debates have been rationalized have often failed to capture the root of the problem.

In fact, both serious philosophical criticism and practical problems, such as with the environment, have called into question the adequacy of 'autonomy or utility' in addressing public policy questions. There is a need to broaden the ethical categories in addressing such areas. It is for this reason that the 'precautionary principle' was developed through a reflection on environmental issues. Public reaction thus needs to be better understood rather than dismissed as 'irrational'. The efforts of the Human Fertilisation and Embryology Authority and of various lobby organisations to persuade the public, without seriously engaging with their concerns, exemplifies that attitude that Bernard Williams once dubbed 'government house' utilitarianism, 'an outlook favouring social arrangements under which a utilitarian elite controls a society in which the majority may not itself share those beliefs' (Williams and Sen 1982: 16). This is problematic from a political and ethical perspective, irrespective of the relative success of this mechanism as a means for securing desired policy. Given the degree of moral pluralism which exists in society, government and ethics bodies have an ethical duty seriously (although not uncritically) to consider the ethical and political views of the population.

9

Worldview perspectives

A worldview is normally understood as a set of beliefs or assumptions through which an individual considers and evaluates the world. As such, every thinking person has a worldview or a set of fundamental commitments that the individual has about the world and the manner in which it works. These are then considered in seeking answers to a set of critical questions (Mitchell et al. 2007: 35) including:

What is the nature of reality, especially what is ultimate reality? (These are questions of metaphysics.)

How can a person know the world? In other words, what do persons know and how do they know what they know? (These are questions of epistemology.)

What is human nature, or what is a human being? (These are questions of anthropology.)

What is right and wrong and on what basis does one make moral decisions? (These are questions of ethics and morality.)

What happens to a person at death?

Where is history going, or what is a person's view of history?

From this perspective, the different worldviews, including religions and belief-systems, concerning human-nonhuman combinations often depend on the manner in which they consider the differences between animal and human species (Modell 2007). Moreover, because human-nonhuman combinations use certain kinds of animal elements, the creation of such entities may cause some concerns to followers of various religions or beliefs, especially if the nonhuman animals used are considered, by them, as impure or having a special status.

For example, in 1993, the Committee on the Ethics of Genetic Modification and Food Use set up by the UK's Ministry of Agriculture, Fisheries and Food published a report examining, amongst other things, the manner in which different religions considered the use of foods containing certain animal genes (Polkinghorne 1993). The report indicated that the Christian community was divided. While some had no objections, many had an uneasy concern, which they found difficult to articulate. This was a feeling shared by many non-Christians and which could be reflected as the 'yuck' factor. The Jewish reaction was more straightforward with the acceptance that 'if it looks like a sheep, then it is a sheep'. Muslims and Hindus, on the other hand, were much more opposed, as were the vegetarians and animal welfare groups.

Interestingly, none of the groups were moved when the committee pointed out that there was effectively no chance of eating the original human gene since it was hugely diluted in the processes of genetic manipulation and by the fact that the gene inserted into the sheep could more correctly be considered as a 'copy-gene' which was completely synthetic (Burke 2006).

For reasons such as these, the following study has examined the respective considerations of the major worldviews concerning human-nonhuman combinations when this was available. However, in some circumstances, only the views relating to xenotransplantation could be determined.

Buddhism

The world's 400 million Buddhists are divided into many sects and schools which have flourished in diverse cultures and lack a clear hierarchy or central source of authority. Furthermore, some Western Buddhists have begun to question the basic presuppositions which in the past have underpinned the more traditional beliefs of Buddhism in Asia. These factors make it difficult to generalize or speak of a 'Buddhist view' on issues such as human-nonhuman experimentation.

An obvious starting point for reflection on this question is the traditional belief in karma and reincarnation. According to this, individuals live not once but many times, and in the course of a long series of lives may cross species barriers such that human beings may be reborn in nonhuman animal form and vice versa. While such an eventuality might be thought to occur only rarely, the belief that it may occur at all clearly presents a radically different perspective to that of the monotheistic Semitic traditions on the relationship between mankind and the nonhuman animal kingdom.

Buddhism, however, also believes that there are important differences between human beings and nonhuman animals, in particular that human beings possess a special dignity and worth by virtue of their distinctive intellectual and moral capacities which uniquely equip them to reach the *summum bonum* of nirvana. The belief that a human being may be reborn in a nonhuman animal species, moreover, does not by itself provide support for the view that it would be desirable to create human-nonhuman combinations.

A universal Buddhist value which would have a strong bearing on any such proposal is that of non-harming (*ahimsa*). In terms of this, any intentional destruction of life is to be avoided, even if it is thought to lead to good ends. Accordingly, experimentation on embryos, whether human or nonhuman (and here the Buddhist objection to the destruction of animal life may be even stronger than that of, say, Christianity) would be immoral.

Where the aim is the creation of life (in the form of chimeras and hybrids) rather than its destruction, the important Buddhist value of compassion (*karuna*) would militate against producing creatures whose destiny would almost certainly be to suffer and to be exploited as a means to others' ends. Instead, compassion would encourage scientists to seek alternative non-destructive and non-harmful means to find cures for those suffering from genetic or other disorders.

In conclusion, the reasons outlined above together with the many biomedical, psychological and other risks associated with the production of human-nonhuman combinations and the complex ethical problems such research would inevitably give rise to, make it difficult to see how it could be supported on the basis of traditional Buddhist values and teachings.

Christianity

Most Christians believe that human beings are made in the image of God which gives them a unique and specific moral value that is greater than that of all other animals. The precise meaning of the 'image of God' (*imago Dei*) has, however, been hotly debated down through the centuries. St. Augustine (354–430) suggested that the Image of God was reflected in the memory, understanding and will of a person. For Augustine, the image of God in the human being is Trinitarian because God is a Trinity of persons (Father, Son and Holy Spirit). Furthermore, this image will be perfected when the human being comes to share in the Trinitarian life of God in heaven. John Calvin, on the other hand, believed that the imago Dei was shown by a person's ethical faculty. The twentieth century theologian Emil Brunner (1889–1966), suggested that the Image of God was characterized in a person's need and capacity for relational love. Others have indicated that it relies on God's decision to form a special relationship with his children (Misselbrook 2004). These qualities are, of course, not mutually contradictory and it is generally possible for Christians to see them as forming a complementary understanding of the Image of God.

Christians also agree that the Image of God is reflected and demonstrated in that God chose a human form, in Jesus Christ, for his incarnate life on earth. This is the essence of the Christian Biblical message, which bears witness to God's intervention in history, through Jesus Christ, to reconcile this corrupted world to himself. But Christians believe that Christ has also called all human beings to follow his example to love God and their neighbours. Thus, many would be extremely concerned about the use or production for research of large numbers of human-nonhuman embryonic entities which/ who could potentially have full moral status and are destined to die. This is because a significant number of Christians would be prepared to give these entities the benefit of the doubt as regards their special or full moral status and consider their destruction as very concerning and/or extremely offensive. Sources relating to the debate in Christianity concerning the moral status of the embryo include, among others, Clarke and Linzey 1988; Waters and Cole-Turner 2000; Jones 2004; Peters, Lebacqz and Bennett 2010. As the Anglican theologian Oliver O'Donovan indicated, if society creates embryos with the intention of exploiting their special status for use in research 'we necessarily stop loving them' (O'Donovan 1985: 65).

Other Christians are prepared to allow the use of human embryos and even the creation of human-nonhuman embryos, if this is intended to relieve human suffering and if it is restricted to a very early stage of development (Seller 2008). Indeed, not all Christians regard the early embryo as a human being made in the image of God. But the Christian tradition overwhelmingly supports the view that the human embryo is specially loved by God and is due some measure of protection. Accordingly, the instrumental use of the human embryo is very difficult to justify by reference to Christian tradition (Jones 2004, Jones 2005).

The Christian physician and ethicist John Wyatt has used the analogy that human beings made in the image of God are 'flawed masterpieces' (Wyatt 2009: 98–9) and that Christians have a duty to correct flaws in the masterpiece to restore it as much as possible to God's intention. Some Christians have argued that this duty can extend to cross-species manipulations, like inserting human genes into bacteria to create human insulin. These are arguably aimed at treating lost functions and 'restoring the masterpiece'.

When it comes to bringing a creature to birth whose identity is ambiguously situated between human and nonhuman, very few Christians would find this acceptable. This is primarily because it would be cruel and unjust to the creature which/who would be born. But Christians might also use the book of Leviticus as the basis for opposing the creation of human-nonhuman embryonic combinations using gametes:

> If a man has sexual relations with an animal, he must be put to death, and you must kill the animal.

> If a woman approaches an animal to have sexual relations with it, kill both the woman and the animal. They must be put to death; their blood will be on their own heads. (Leviticus 20:15–16, see also Leviticus 18:23)

Arguably, the reason that bestiality is condemned here is that the human person may be considered as having completely given himself or herself over to the animal through a sexual act. The person would be regarded as having offered his or her whole humanity or even his or her image of God within himself or herself to the animal, thereby undermining the whole concept of what is special in humankind and which reflects God.

Many Christians also accept that human beings have been given stewardship over animals and are permitted to use them to benefit humanity. As is written in Genesis 1:26:

> Then God said, Let us make man in our image, in our likeness, and let them rule over the fish of the sea and the birds of the air, over the livestock, over all the earth, and over all the creatures that move along the ground.

In this respect, serious Christian concerns would involve the violation of the divinely created order (HCSTC 2007, HC272-II, Memorandum 53: Submission from the Church of Scotland and the Church and Society Council) and the potential suffering of experimental animals. Moreover, most Christians believe that God designed procreation so that plants, animals and humans always reproduce after their own kind or seed. As indicated in the book of Genesis:

> So God created the great creatures of the sea and every living and moving thing with which the water teems, according to their kinds, and every winged bird according to its kind. And God saw that it was good (Genesis 1:21 see also Genesis 1:11–12 and 24–5).

Therefore, from a biblical perspective, species integrity is ultimately defined by God rather than simply by arbitrary forces. Accordingly, the fusion of, for example, human-nonhuman genomes may be perceived as running counter to the sacredness of human life and humanity created in the image of God (Jones 2003; Cobbe 2007).

In 1987, The Vatican's Congregation of the Doctrine of the Faith indicated that any attempt of fertilization between human and animal gametes and the gestation of human embryos in the uterus of animals were contrary to the concept of human dignity (CDF 1987, paragraph I: 6). Moreover, in 2007 Bishop Elio Sgreccia, president of the Vatican's Pontifical Academy for Life, was cited as saying that: 'The creation of an animal-human hybrid embryo is a step over a barrier which has until now been prohibited by all in biotechnology because human dignity is compromised, affronted, and that monstrosities could then be created through these procedures'. Adding, 'the creation of an animal-human being represents one of the most serious violations of a barrier in nature giving rise to a comprehensive moral condemnation' (Nau 2007).

In December 2008, the Congregation for the Doctrine of the Faith of the Holy See issued a document which contained explicit reference to the reprogramming of animal oocytes by the nuclei of human somatic cells:

> From the ethical standpoint, such procedures represent an offense against the dignity of human beings on account of the admixture of human and animal genetic elements capable of disrupting the specific identity of man. The possible use of the stem cells, taken from these embryos, may also involve additional health risks, as yet unknown, due to the presence of animal genetic material in their cytoplasm. To consciously expose a human being to such risks is morally and ethically unacceptable. (CDF 2008, paragraph 33: 19)

Bioethicist, Nicholas Tonti-Filippini and others have also argued that the creation of human-nonhuman embryos is an offence against a Christian view of human generation:

> It seems to us that when a scientist fragments the human genome and adds parts of it to an animal genome in the formation of a hybrid zygote, he or she has begun to confuse the identity of what is or is not human and what or who is or is not made in the image and likeness of God, and does or does not count as my neighbour'. (Tonti-Filippini et al. 2006: 704)

It is on that ground that the authors hold that such a project represents a failure to respect the sacredness of the human genome and the sacredness of human generation. Reflecting on the mystery of the Incarnation, these same authors note that,

> it is the human genome that the Second Person of the Blessed Trinity has taken to himself. This decisive event fundamentally alters the way in which we should respect the sacredness of the generative capacity of the human genome when it is used to form a zygote. (Tonti-Filippini et al. 2006: 704)

In summary, while recognizing the possible usefulness of interspecies research in some circumstances, many Christians would believe that if the moral status of a proposed human-nonhuman combination cannot be determined without creating such an entity, then that in itself should be a sufficient argument against creating such an entity. (This is also the position taken by the Church of Scotland General Assembly in its 2006 Deliverances which 'Oppose the creation for research or therapy of parthenogenetic human embryos, animal-human hybrid or chimeric embryos, or human embryos that have been deliberately made non-viable'.)

Hinduism

Hinduism originated over 3,000 years ago and claims to have many founders, teachers and prophets who claim first-hand experience of God. Hindus promote the idea of spirituality as a principle rather than a personality and call this Brahman. Because Hinduism is a term that includes many different although related religious ideas, there is no clear single Hindu view on the right way to treat animals or on the creation of human-nonhuman embryonic combinations. The doctrine of *ahimsa*, however, leads Hindus to treat animals well and most are vegetarian.

Generally, Hindus would believe that nonhuman animals are inferior to human beings. But the cow is greatly revered by Hindus and is regarded as sacred. Killing cows is banned in India and no Hindu would eat any cow product. Interestingly, some Hindu gods have human-nonhuman characteristics. Indeed, Ganesh has the head of an elephant and Hanuman has monkey body parts.

Most Hindus do not believe in procedures such as xenotransplantation since the body must remain whole to pass into the next life. However, some Hindus would be willing to make exceptions while others would recognize that xenotransplantation is a matter for individual choice. In this respect pig and sheep organs would be acceptable.

Islam

The contents of this section are further developed in Panjwani and Panjwani 2010. Whether in an Islamic or non-Islamic framework, one could obviously tackle the creation of human-nonhuman combinations from any number of perspectives – moral, legal, political, ethical, spiritual, scriptural and so on. What will be specifically addressed in the following paragraphs, however, are the philosophical and metaphysical perspectives pertaining to the concepts of *life, soul* and *spirit*, within an Islamic context, and the significance they have in human-nonhuman combinations and creations.

Perhaps the starting point is to consider what is meant by 'life'. According to Hasan al-Basri, an eighth-century theologian, life is an expression of knowledge and ability (Hilli: 12). In other words, life is not an independent existence but rather a description of a particular existence possessing certain qualities. This restricts life to those subjects who possess these character-istics, which at first instance, would be human beings. The Qur'an indicates 'We have shaped mankind in the best mould' (Q. 95:4), 'And surely We have honoured the children of Adam and We carry them ...' (Q. 17:70). Yet, if animals were also considered as possessing a level of knowledge and ability, then they would also come under this definition.

Rida al-Sadr, however, defines life as an expression of mobility and repro-duction (Sadr 1986: 271). This would not only include human beings and animals but plants as well. Hence, if life is seen from a rational perspective, then animals and humans become much closer in the attributes that define them.

Life is, therefore, an expression of certain attributes (rationally derived) displayed by existences at particular levels of their existence.

But deliberating on the concept of 'life' may still not give an individual the metaphysical enquiry which is necessary to analyse the essence of a being. For this, one would have to investigate the concept of the 'soul'. If life is explained in terms of motion, ability, knowledge and perception, then the soul is defined in terms of a unique level of self-consciousness or self-awareness that gives a being its distinct essence and identity. This can be explained by referring to the first Shi'a Imam, Ali ibn Abi Talib (598–661), who indicated that, 'The souls are four, the growing vegetative soul, the sensually perceiving animal soul, the rational human soul and the Godly soul' (Kashani 1980: 267).

This indicates that human beings are characterized by the third kind of soul but in turn possess animal and vegetative souls in a manner that seems to be progressive and evolutionary. But can it be said that animals are subject to such divisions within their souls, if they possess souls? The Qur'an explains the Godly awareness of animals:

To Solomon We inspired the (right) understanding of the matter: to each (of them) We gave Judgement and Knowledge; it was Our power that made the hills and the birds celebrate Our praises, with David: it was We Who did (all these things). (Q. 21:79)

Furthermore, is it correct to say that the faculties of the soul, as above, are perhaps more progressive in human beings? What would be the result if certain organs were transplanted from animals into human beings or vice versa? Would it affect the entity and/ or growth of the soul, whether it is separable or inseparable from the being? These questions, which of course require extensive deliberations, perhaps point to one major theme, as indicated: the self-consciousness of an individual. To what extent will the self-consciousness of an animal or human being be affected? To give us some insight into this question, we can refer to Abd Allah ibn Sina (commonly known as Ibn Sina or Avicenna (c. 980–1037)) and Sadr al-Din Shirazi (commonly known as Mulla Sadra (c. 1571–1636)).

Concerning the soul's relationship with matter, Ibn Sina believes that the universal soul parts with a particular faculty of the soul as the material subject is ready to receive it in the process of its growth. Furthermore, it is the soul that is responsible for creating motion within matter due to its love for re-ascending to God (Nasr 1993: 207).

For Sadra, the soul is a power or a form that works on matter not directly but through other forms. This would mean that the human soul is a power that operates through the vegetative and the animal souls. And, the human soul is not something that is separate from the body and enters within it, rather it is something that is produced at a particular stage of the development of the body (Rahman 1975: 199).

Therefore, if human-nonhuman creations were to occur, it is necessary to consider a being's development as a particular entity (for example, an animal is subject to its common animalistic developments) and as itself (for example, an animal's own personal identity). If the soul occurs when the material subject is ready to receive it or at a particular stage and yet transplants have occurred, would it hinder the self-consciousness of the being, since matter and self-consciousness can be inherently connected? Is it not possible to deliberate that in nature as well as in scriptures, the reality that God has created groups with particular characteristics and functions, indicates detail and design? 'Verily, all things have We created in proportion and measure' (Q. 54:49). If these are disrupted, then each being fails to complete itself, in terms of developing and expressing its self-consciousness.

So, the soul is a unique level of self-awareness as well as the source of the attributes of life that the being displays. The soul allows the distinct animalistic and intellectual faculties to develop in a being which then contributes to the growth and identity of that being.

In considering the development of a being, another relevant concept is the 'spirit' (*rūh*). According to Islamic philosophical and mystical traditions, the spirit is seen as among the primary emanations to come from the Divine essence (Rahamni 1997: 275). It has various meanings. It is described as a creation in itself but above the rank of angels. It can also be a divine aid assisting God's servants. At other times, it is used in the sense of a life giving agency of God. It is considered as something issued from the command of God of which God has given very little knowledge (Q.17: 75). Finally it is used as something that descends with the command (*amr*) of God to the earth with the angels and ascends to the heavens (Q.97: 4) (Q.70: 4). It is also possible to characterize the spirit as the cause for the emergence of the human soul.

All of the above would designate the spirit at the rank of humans and above. Perhaps at the very least, the spirit plays a part in the emergence of the human soul and its subsequent evolution through its process of actuality. Therefore, if a being's journey is to complete itself towards God and this requires a stable development of self-consciousness with the issuance of the Divine spirit, any deliberations on human-nonhuman combinations or creations should consider this aspect of a being as well.

In conclusion, whether or not one is a religious believer or spiritual person, it is still necessary to consider the dimension and growth of self-consciousness in a being. Human-nonhuman combinations could be beneficial when the overall aim is to maintain the unique and natural course of self-consciousness in all beings, such as in the therapeutic realm. Beyond that, caution must be exercised before proceeding with bolder experiments. Perhaps the first step is to consider the aspect of self-consciousness of a being seriously. Tampering with this may affect the beautiful journey of that very being to complete itself, with its own self-awareness.

Judaism

In Judaism the most important principle in medicine is to save human life. This even over-rides other considerations such as the prohibition on the consumption of pork. In Judaism, concerns would also be raised about safety, the suffering of the animals and over interfering with the order of Nature.

For many in Judaism, using spare embryos from fertility treatment before the fortieth day after conception is permissible for appropriate reasons since this is the stage where human life becomes morally important. But diverging views do exist such as the Jewish commentator Eric Cohen who is Director of the Bioethics and American Democracy Program at the Ethics and Public Policy Centre in Washington, D.C. Indeed, he believes that because (1) medical advances have enabled embryos to live outside the human body and (2) science has indicated that the 40-day stage after conception is not a significant time in human development, this has put humanity in a situation unanticipated by religious tradition. As an alternative, Cohen believes that human life should be respected from conception onwards and warned of the dangers of defining a class of human beings as unworthy of life from a human and not God-centred perspective (Powell 2007).

Other members of the Jewish faith believe that it is only in the creation of an embryo more specifically for experimentation, and not for reproduction, that a more fundamental contradiction of the status of the human embryo exists.

Little has been written from the Jewish perspective on the issue of human-nonhuman embryonic combinations. The creation of some forms of human-nonhuman interspecies embryos may be authorized in Jewish law in the context of eventually saving lives and protecting or restoring health, though the transformation of one species into another would be prohibited (Shapira 2009: 765).

In responding to the UK House of Commons consultation on such combinations in 2007, the Office of the UK Chief Rabbi indicated that at 'this stage' it had 'no objections to the creation of animal/ human chimera or hybrid embryos for research purposes' but nevertheless hoped that the Committee would 'continue to consult, as the implications become clearer' (HCSTC 2007, HC272-II, Memorandum 62: Submission from office of the Chief Rabbi, Sir Jonathan Sacks).

Secular worldviews

Amongst those who tend towards more secular philosophies, no single position is possible because of the number of different views relating to the creation and use of human-nonhuman embryonic combinations. There is also a question as to what is meant by a secular philosophy. Does secular equate with atheist? In the sense in which the term is used here, a 'secular worldview' is one that does not presuppose or require a commitment to religious belief though other kinds of beliefs may be present. It may, in fact be combined with a religious vision. There are Jewish Marxists, Islamic feminists, and Christian Humanists (though this last term has been used for more than one worldview). It should also be noted that these secular world-views may well be combined to a lesser or greater extent. Nevertheless, what characterizes these worldviews as secular is that they are able to sustain an ethical approach to life without religious commitment.

Evolutionary thought

Within the contemporary world it would be hard to underestimate the influence of evolutionary thought and in particular the theory of the origin of species by natural selection as originally put forward by Charles Darwin (1809–1882). Evolution proposes to connect human life to that of other nonhuman animals. It also seeks to explain the nature and existence of human life without recourse to religion (which does not imply that a scientific account of evolution must be incompatible with belief in God).

In relation to human-nonhuman combinations, evolutionary theory may be taken to suggest a view of species as being provisional and representing a fluid collection of individuals rather than a set of stable 'natures'. This in turn may undermine a doctrine of the special status of human nature or of the human species. Hence the overall influence of evolution tends to make human-nonhuman combinations less threatening. Indeed they may be welcomed precisely as living proof of the evolutionary principle.

Nevertheless, while evolutionary thought raises important concerns about the use of the term 'human species', the discussion of ethics is precisely a discussion of human action, that is, with the needs, passions and actions of human beings. The recognition that human beings may have evolved from other species need not undermine the recognition of their common humanity, which is the beginning of ethics.

Materialism

Evolutionary thought is often combined with materialism, the doctrine that the only thing that can truly be said to exist is matter. In other words, all things are composed of matter and all phenomena are the result of material interactions. Modern forms of materialism tend to be 'reductionist' in the sense of holding that matter is to be understood by reference to the behaviour of the smallest parts. One well known example of biological reductionism is the concept of the 'selfish gene' developed by Richard Dawkins (1976). Reductionism about biology tends to undermine concepts such as human nature. It is frequently found with utilitarianism in moral theory, as this also analyses human action in relation to 'units' of happiness (Mill 1987; Williams 1982). From this perspective, human-nonhuman combinations should be permitted if they increase the amount of net happiness (measured in utilitarian terms).

There are, however, forms of materialism that are not reductionist and which allow for 'emergent properties' of matter as bodies become more complex. Such an approach would support a richer view of human life and a humanism that had space for concepts such as freedom and human dignity.

Humanism

Humanism is a broad category of ethical philosophies that affirm the dignity and worth of all people, and is based on the ability to determine right or wrong by appeal to universal human qualities, such as rationalism. Humanism entails a commitment to the search for truth and morality through human means in support of human interests.

Though the term once referred primarily to a movement within religious traditions, in the contemporary world it generally refers to secular philosophies. Many 'humanist' organisations, such as the British Humanist Association, do not have any particular ethical objections to creating human-nonhuman embryonic combinations for research purposes, provided that their use is regulated in the same way as the use of human embryos for research. Secular humanism has its roots in the eighteenth century enlightenment and places a high value on reason as the basis for morality. Hence, while they might accept that some people might find the idea distasteful, they would consider this an irrational response given that the embryos would be destroyed at a very early stage.

Many humanists would probably welcome the creation of such embryos, as they would reduce the shortage of human embryos for research and so allow progress to be made in various kinds of medical research, which would benefit society (information provided by Ms H. Stinson of the British Humanist Association, 6 March 2007). Humanism also places a high value on choice and autonomy and would generally support freedom of research unless there was a clear reason to restrict it.

Nevertheless, while many humanists, particularly in the English speaking world, take a pragmatic approach to this question, there are some, particularly but not only in continental Europe, who have questioned whether some biotechnological innovations (such as cloning, germline genetic engineering and crossing the species barrier) might be incompatible with respect for human dignity. The concept of human dignity has been foundational for many accounts of human rights. It is a matter for further reflection whether this profoundly humanist concept (in the broadest sense of this word) is compatible with creating human-nonhuman combinations.

Pragmatism

Pragmatism is a philosophical movement that includes those who claim that an ideology or proposition is true if it works satisfactorily; that the meaning of a proposition is to be found in the practical consequences of accepting it; and that unpractical ideas are to be rejected. Pragmatism does not consider any basic difference between practical and theoretical knowledge nor any fundamental difference between facts and values. Values then become suppositions of what is good in practice. Pragmatist ethics does not consider any moral attribute beyond those that are important for human beings and is, therefore, very similar to humanist ethics. In this context, pragmatists would suggest that the creation of human-nonhuman embryonic combinations may be acceptable if they can be practically applied in biomedicine in useful contexts.

Marxism

Marxism, which was heavily influenced by Karl Marx (1818–1883), has a materialist concept of history representing an alienation of the working class which is separated from the product of its work while often being exploited by a rich economic minority. Society, therefore, has to change to enable workers to take pride or personal satisfaction from their labours. Creativity and the possibility of benefitting from work are to be seen as a way of becoming fulfilled as a person. The important feature of Marxism with respect to ethics is the central aspect of personal development. In addition, it sees the moral self as creating something which can be considered as positive, and not just responding to circumstances.

Marxism has also been closely associated with an atheistic worldview in which religion was considered as limiting the working classes in their liberation from exploitation. It was because of such an atheistic setting that Lev Fridrichson, a representative of the Soviet Commissariat of Agriculture, expressed the hope in 1924 that the creation of living human-chimpanzee hybrids 'should become a decisive blow to the religious teachings, and may be aptly used in our propaganda and in our struggle for the liberation of working people from the power of the Church' (Friedrichson to Aleksandr Tsyurupa, deputy chairman of Soviet government, 20 September 1924, in Rossiianov 2002: 286).

On the other hand, Marxists would also ask about the money, time and social prestige that are invested in human-nonhuman research projects. They would take note of the links between universities and biotech businesses, and the debates over patenting life and intellectual property issues. Thus, though they would be unlikely to object to such experiments in principle, Marxists would be less likely to take claims of prospective benefits from such research at face value but would ask whose interests are served by these claims and whether these developments are likely to address the health needs of most people.

Feminism

Feminist ethics usually traces its roots back to the eighteenth century. Like Marxists, feminists ask about the interests of those who determine traditional values and roles, and asks about the impact of those values on roles of those who have been excluded from this decision making process. However, while Marxism concerns economic power relations, feminism concerns the power relations between men and women. Feminist commentators indicated that many ethical positions on matters as diverse as marriage, work, education and healthcare had not been reached through purely rational arguments, but had been based on an impoverished vision which has neglected women's voices and from which women have disproportionately suffered. While male philosophers had lauded the autonomous free (male) individual they had typically left unacknowledged the servant classes on whom they relied and, more particularly the work of women as educators and carers.

In the twentieth century, feminist ethics brought with it a new awareness of womanhood, sexuality, contraception and new forms of reproduction. There are differences here between those feminists who promote ideas such as 'reproductive autonomy' as empowering women, and other feminists who are critical of the concept of autonomy itself as requiring a more radical feminist critique. There are also a variety of views on the creation of human-nonhuman combinations. Many feminists are uneasy about recognition of the moral status of the embryo *in vitro* because of the implications of this for the reproductive autonomy of women (Baylis and Mcleod 2006 criticizing Holland 2001). Nevertheless, there has been a vigorous feminist critique of the embryo research industry, not because of the impact on the embryo, but because of the impact on women, especially in relation to the use of human eggs. With respect to human-nonhuman combinations, the development of cybrids was promoted precisely as reducing the negative impact on the technology on women, but some feminists have voiced scepticism about this ostensive motive (Baylis 2008).

Environmentalism

The environmentalist or Green movement has its roots in nineteenth-century thought but came to the fore in the late twentieth century as the negative impact of human behaviour on the environment became more apparent. It has become a worldview which has provided a new way to understand the place of human beings in the world and the ethical challenges that face humanity. Keenly aware of the potentially harmful effects of human activity, environmentalists tend to be sceptical of novel technologies and apply a precautionary principle in cases where it is difficult to assess the future impact. Related to environmentalism is the issue of animal welfare (or 'animal rights') in agriculture and in medical research. However, these are distinct concerns and might pull in opposite directions (as when animal-welfare activists release mink into the environment, see Atkins 2000).

Environmentalists will be concerned about potential harms that might be caused by crossing the species barrier, both potential harms to human health from new diseases and potential effects on other species. They will also place a high value on animal welfare and will generally resist new forms of technology that would involve further use of animals, seeking to promote alternatives to this. In Germany the Green movement has been active in its opposition to biotechnologies that involve experimentation on human embryos, and would certainly be opposed to admixed embryo research. In the United Kingdom, the focus of Green concern has been with agricultural biotechnology (and the GM food debate) rather than with biomedical technologies. Nevertheless, in general it could safely be said that the environmental movement would be more sceptical about the alleged benefits from creating human-nonhuman combinations, and more concerned about potential harms.

10

Ethical perspectives

Ethical analysis studies values and customs of a person or group including concepts such as right and wrong as well as responsibility. Ultimately, however, ethical debates are likely to be based on worldviews, i.e. fundamental beliefs and value judgements, which cannot be 'proved' right or wrong by a single simple test or argument. This does not mean that good arguments are futile in rational discourse; it is just that such profound judgements come not from abstract arguments but from a convergence of reasoning. As the philosopher Wittgenstein (1889–1951) said, 'light dawns gradually over the whole' (Wittgenstein 1969: 21).

Many people would have grave ethical concerns about the creation of human-nonhuman entities. In crossing this barrier, the general understanding of what it means to be human would no longer be clear-cut. Indeed, any ethical appraisal of crossing this barrier should ultimately answer the question of whether combining human and nonhuman elements changes the identity and the rich meaning of what is understood to be a human or nonhuman animal. And if modifying a human or nonhuman animal body is being considered, then questions relating to the acceptable limit of such modifications may be posed (Pontifical Academy for Life 2001).

In order to seek to clarify the different ethical issues relating to a procedure, such as the creation of human-nonhuman combinations, it may be useful to seek to classify the different arguments which considered such a procedure as either intrinsically wrong, in itself, and arguments which view it as extrinsically wrong because of its consequences. This important distinction can be applied to a large number of moral issues and can often be of assistance in identifying the reasons behind moral concerns (Straughan 1999: 11).

Extrinsic concerns

Extrinsic reasons for accepting or opposing an action or process are related to the possible consequences of that action or process. But these consequences and whether they are good or bad are sometimes very unclear and agreement is difficult to reach. Moreover, there are always a number of consequences to any activity, often occurring at different times or stages, which have to be weighed and compared against one another, and cannot be based on a straightforward objective assessment (Straughan 1999: 17).

For example, some people argue that the value of research depends on its utility. Arguments within this school of thought involve weighing up the potential value of a technology (for instance, the possible future development of an improved treatment for some disease), against the reasons for not doing so. Arguments within this category may also take into account the 'slippery slope'. In other words, a number of individuals may not object to the creation of human-nonhuman embryonic combinations for their immediate purpose (to produce embryonic stem cell lines), but they may, nonetheless, have (ultimately overriding) concerns about what might follow. In short, that allowing scientists to create human-nonhuman embryonic combinations may eventually result in the birth of such creations (HCSTC 2007, paragraph 36).

Risks of biological developmental problems

One of the first perspectives to consider in relation to the 'extrinsic concerns' is the very real risk of dysfunctional biological development in human-nonhuman embryos. In this regard, it should be recalled that when Dolly the sheep was created, in 1997, the first reproductive cloning experiment ever reported, it took 277 nuclear fusions to produce eight embryos, which only yielded one viable lamb (Wilmut et al. 1997).

A far greater number of pre- and post-natal developmental biological problems would occur in the creation of human-nonhuman combinations using the same or other approaches, especially at the embryonic level. This would then probably give rise to serious concerns for many commentators relating to the creation of large numbers of entities of human or near-human moral status which are destined to be destroyed through 'wastage'.

Risks of creating new diseases

Another important concern is that many animals may harbour microbiological and other entities for which they are healthy carriers in their organs, cells and genome, because they have developed protective mechanisms to resist them. Some of these entities, however, are capable of crossing the species barrier and developing in the host. Unfortunately, the appearance of new diseases through the crossing of the species barrier is not a myth. Any infectious disease that may be transmitted from other animals, both wild and domestic, to humans or from humans to animals is called a zoonosis. Prion diseases, such as Creutzfeldt-Jakob Disease, can be contracted by humans through eating material from animals infected with the bovine form of the disease. Avian Flu is adapted to birds, but this does not mean another species cannot catch the flu; nor does it mean it cannot adapt to human beings. Finally, the human immunodeficiency virus (HIV) is very probably of simian origin in which the nonhuman animal has ceased to play any part in the pandemic which it has caused (French National Consultative Bioethics Committee 1999).

In xenotransplantation using animals such as pigs, there is also the potential for serious infectious diseases to spread from the donor animal. Some such infectious agents are porcine endogenous retroviruses (PERVs) which are viruses within pigs to which they are immune, but which may infect human cells. Thus, some recipients of pig neural cell transplants have had to agree to never donate blood, take frequent blood tests and use safe sex methods for the rest of their lives because of the risk of spreading such viruses.

In short, creating human-nonhuman combinations may bring about a period of uncertainty (knowing hazards but not the probability relating to their occurrence) and even ignorance (hazards occurring that one did not even envisage) as to the possibility of spreading new diseases. Physicians and biologists have to ask whether the infectious risk is sufficiently serious for it to ever be ethical to allow humankind to run the risk of devastating and uncontrollable pandemics since human-nonhuman combinations will never concern more than a limited group of procedures (Butler 1998).

Psychological risks

In addition to the biological concerns, substantial psychological risks would almost certainly be encountered in the eventual, but very improbable, situation that some human-nonhuman combinations were actually born and capable of self-awareness. These would, for example, be related to their awareness of their unique human and nonhuman origins and identity. A very real risk would exist relating to the manner in which the individual would perceive himself or herself.

Concerns would also exist as to the manner in which society would perceive, accept and relate to human-nonhuman combinations. A question could be asked whether the general public would, for instance, be tempted to discriminate against any being that was not seen as having a human identity.

Risks to women

The history of attempts at human-nonhuman combinations has frequently involved the exploitation of women. This is seen perhaps most graphically in the proposal previously mentioned that women be impregnated without their knowledge with sperm from a nonhuman primate. But even if the woman consented to the procedure, the medical risks to a woman in gestating a human-nonhuman foetus make this ethically unacceptable. Without consent it would have been doubly abusive.

The question of safety is an extrinsic concern in that it is possible, at least in principle, that the technology may become safe in the future. Nevertheless, for the foreseeable future it is a pressing ethical question.

As the gestation by a woman of a human-nonhuman foetus is an example of unacceptable risks, so women may also be abused by extracting their eggs for research purposes. The collection of human eggs is an invasive procedure and it is typically preceded by the hyper-stimulation of the ovaries so that a greater number of eggs can be 'harvested'. Both these procedures carry risk, but they are risks that many women freely undergo in search of a child. However, when the woman is not receiving fertility treatment for her own sake then there is a question as to whether these risks are warranted. The ethical concerns are even greater when a woman is paid for her eggs or receives free fertility treatment in exchange for the diversion of some of her eggs to research use. It is noticeable that the scandal surrounding the South Korean scientist Hwang Woo-Suk began with just such an exploitation of women.

Some of those who argued for the licensing of research on human-nonhuman cybrids did so on the basis that this would help protect women from this kind of exploitation. If embryos are created from cow eggs then human eggs need not be used. Thus the argument for a specific kind of human-nonhuman combination was alleged to be its ethical benefit as an alternative to taking eggs from women. Nevertheless, feminist critics have been sceptical of this sudden conversion to the consideration of the protection of women egg donors. The only research centre in the United Kingdom to begin research in the area, the Newcastle Life Centre, did not renounce the use of human eggs in research. Indeed they continued a policy of paying for fertility treatment as an indirect way of paying women for their eggs. This is hardly surprising because the project of human-nonhuman cybrid research takes its place within the larger project of therapeutic cloning. The aim of the research was to improve knowledge and efficiency by using cow eggs with the intention of then using women's eggs in therapy.

The phenomenon of discussing the ethical implications of embryo research without considering the impact on women was summed up by Donna Dickenson as the 'the lady vanishes' (Dickenson 2006). This is just as much of a problem with cybrid research as it is with other forms of embryo research. In principle the use of cybrids is an alternative to using human eggs, but in fact it takes its place in a larger project in which women are used but this is not considered ethically significant (Baylis 2008). It is perhaps for this reason that President Obama restricted funding to research that created human embryos whether from human or nonhuman eggs. These are not so much alternatives as elements of a common project.

Animal welfare and animal rights

Before examining whether and to what extent certain nonhuman animals should be accorded welfare and rights it is necessary to consider the manner in which these animals may be attributed moral status or importance. In this regard, there are generally three different ways of considering this status (Nuffield Council on Bioethics 2005: 38):

> That there is a categorical moral difference between the moral status of a human being and that of a nonhuman being. This is based on the belief that there is something very special in human beings, as a *Homo sapiens*, that nonhuman beings lack.

> That no clear difference exists between human and nonhuman animals but that a continuum of different degrees may exist between living beings based, for example, on their neurological capabilities. In this sense it is considered that human beings are at the top of a sliding scale.

> That a classification between different species is not sufficient to distinguish between, or attribute, varying moral statuses or importance.

For example, it could be argued that all animals that can suffer may be considered as being in the same category of moral importance. As the philosopher Jeremy Bentham (1748–1832), who influenced a number of his colleagues in this modern day, indicated:

> a full-grown horse or dog is beyond comparison a more rational, as well as a more conversable animal, than an infant of a day, or a week, or even a

month, old. But suppose the case were otherwise, what would it avail? the question is not, Can they reason? nor, Can they talk? but, Can they suffer?'. (Bentham 1823, Ch. 17, Section IV, Note 122)

For some individuals, animal welfare is one of the only significant ethical issues raised by human-nonhuman combinations such as transgenesis (Sandoe and Holtug 1993). More generally, however, commentators acknowledge that animal welfare and animal rights are among a number of aspects that have to be considered together.

With respect to human-nonhuman combinations, the CHIMBRID project (funded by the European Community to examine legal and ethical approaches to the creation of the human-nonhuman combinations) suggested in its 2008 report that there are six moral considerations for animal ethics (Taupitz 2008):[1]

1 Pain and suffering;

2 Substitutability/replacement;

3 Animal quality of life;

4 Treatment of animals appropriate to their species;

5 Species integrity; and

6 Debasement or adulteration of [sentient] life.

If at least one of these different features exists, it is suggested that certain constraints or limits on the manner in which the living being may be treated should be respected.

This CHIMBRID report also indicated that though the replacement of animal experiments is one of the standard criteria in animal ethics, the suffering and quality of life of animals can generally be weighed against the benefits the experiments can produce.

One ethical argument in favour of research on interspecies embryos is that these could facilitate the development of cell-based technologies which would in turn diminish the need for animal experimentation. However, even

---

[1] These were generally similar to different features suggested by the Nuffield Council on Bioethics in its 2005 report entitled 'The Ethics of Research Involving Animals' that may qualify nonhuman animal (but also human beings) as moral subjects. Moral subjects are defined in the report as 'beings whose features should be taken into account in the behaviour of moral agents'. In this regard, moral agents are 'beings that are able to behave in a moral way and are liable to moral criticism for any failure to do so'. The Nuffield Council on Bioethics suggested five features that may qualify living beings as moral subjects, namely (1) sentience, (2) higher cognitive capacities, (3) the capacity to flourish, (4) sociability, and (5) the possession of a life (Nuffield Council on Bioethics 2005: 39-41).

with cell-based technologies there remains a need for animal experiments to test for safety before introducing the technology to human beings. Thus, even if human-nonhuman combinations are not allowed to develop past the embryonic stage (and therefore are incapable of suffering) the research programme will undoubtedly involve experiments on animals whose welfare is a legitimate concern. If the combinations are born then this raises welfare issues, especially if they are also subject to the risks of developmental abnormalities.

The report further identified specific issues when experiments are performed with 'higher' animals, such as primates, because of their greater degree of sensitivity. There is real concern that they may actually develop and exhibit features similar to humans. This raises deep questions of whether this is intrinsically wrong, a question addressed later on. Given this combination of extrinsic and intrinsic concerns, many commentators consider a general prohibition of experiments involving nonhuman primates to be appropriate.

Sentience

Sentience, or the capacity to experience pain and pleasure, has increasingly been proposed as an overriding qualification for the possession of rights. From this perspective, it has been suggested that practices that combine human and nonhuman tissue, in common with practices that enrol animals into human activity such as farming, animal biotechnology, medical research, animal circuses, may only cast nonhuman animals as objects rather than subjects. In other words, as merely the means through which one produces knowledge, technologies, capital, better humans etc. rather than sentient beings having things done to them and experiencing invasive procedures, such as suffering and death.

Treating nonhuman animals as experiencing subjects means taking into consideration more than their welfare (the idea that under human control, an animal should not suffer unnecessarily). Attention to welfare alone is usually only undertaken in the context of reduced or 'avoidable suffering'. An animal might still suffer if that suffering is thought to be unavoidable and counter-balanced by the benefits of a particular research programme.

In addition to animal welfare, the claim for animal rights is to state that animals ought not to be used or regarded as human objects or possessions because they have moral status and basic interests, such as in avoiding suffering, that are shared or otherwise on an equal footing with those of humans. To believe otherwise, it is suggested, would be an example of 'speciesism': unjustified discrimination on the grounds that a particular animal or group of animals is not human (Singer 1976). From this view of animal rights, attention to welfare alone is not enough, for any suffering is inherently wrong as a first principle.

It is proposed, moreover, that animal biotechnologies encourage a reductionist view of animals. In this way, breaking up animals into constituent biological parts (genes, cells, gametes, organs etc.), whose relationship to the organism in which they reside is wholly contingent, threatens the integrity as well as the autonomy of animals (Holland 1990; Bowring 2003; Hauskeller 2005). But this argument is not altogether new in biotechnology. Sixty years ago Collingwood commented that for a cattle-breeder, an improved form is one better suited to that breeder's interests, and these are not identical to that of the cattle (Collingwood 1960 (originally 1945)). A claim such that 'it is morally indefensible to knowingly inflict suffering on sentient animals' must be qualified in at least two respects: first, by the inclusion of the term 'innocent animals', since otherwise the punishment of criminals (who are also sentient animals) would be indefensible, and second by the phrase 'other than in their own interests', since otherwise painful veterinary treatment would be indefensible (Animal Procedures Committee 2003). The inclusion of these and other necessary qualifications weakens the strong argument for the avoidance of suffering.

Some commentators have also argued that the very notion of 'animal rights' is absurd. The philosopher Roger Scruton argues that if animals have rights, then they also must have duties which would make almost all of them habitual law-breakers requiring punishment (Scruton 1996). But even if most animals have no duties towards human beings, this does not imply that humans have no duties towards animals – since it is humanity that exploits and manipulates the animal kingdom for its own benefit (Hartman and Williams 1993).

Nonetheless, many scientists argue that research on animals is necessary before any products are given to human beings. When this is the case, it is accepted that whenever possible, research on non-sentient animals should be preferred over those which are more sentient. Thus the creation, gestation and breeding of an animal which has a potential for sentiency to be significantly increased should be considered and assessed in a very careful manner (Hyun et al. 2007).

As a means of summarising the main features relating to the sentiency of animals, the Banner Committee in the UK produced a report for Government Ministers in 1995 on the ethical implications of emerging technologies in the breeding of farm animals (Banner 1995). This recognized that genetic modifications might result in welfare problems which may not always be immediately apparent. Because of these features, the Committee produced three general principles for future practice:

1 Harms of a certain degree and kind ought under no circumstances to be inflicted on an animal.

2 Any harm to an animal, even if not absolutely impermissible, nonetheless requires justification and must be outweighed by the good which is realistically sought.

3 Any harm which is not absolutely prohibited by the first principle, and is considered justified in the light of the second, ought to be minimized as far as is reasonably possible (Straughan 1999: 21)

Animal telos

The concept of telos, sometimes also termed 'intrinsic value' or 'integrity' of a being, goes back to Aristotle, but has been recently revived by certain commentators (Rollin 1986; Verhoog 1992; Vorstenbosch 1993). It has been aptly summed up as 'the pigness of a pig', the sum total of an animal's genetic potentialities, whether realized or not (De Pomerai 1997).The resulting argument is that by interfering with an animal's inherent characteristics, one may be changing what that animal 'is', its 'natural' form of life, its purposes and ends. As such, it may arguably also be considered under intrinsic ethical concerns which will be examined in the next section.

The principle of telos or intrinsic value can be applied to any living organism, whether sentient or not. This does not diminish the significance of modifying telos through human-nonhuman combinations, such as transgenesis, in sentient animals, although it does make the boundaries of ethical concern more fluid and the issue becomes more difficult to define.

Modifying telos is not necessarily the same as violating telos and it is possible to question whether all genetic manipulations must necessarily compromise telos. It may be conceivable for a genetically modified animal to live a fully normal life, giving free rein to its intrinsic nature and preferences. If the animals, whether genetically modified or not, are able to fulfil their needs and instincts then it is possible to assume that their telos is being respected and that they are capable of flourishing (Nuffield Council on Bioethics 2005: 44). Despite its lack of conceptual clarity, however, the question of telos deserves serious attention. In most cases it is obvious that violations of animal telos will also result in some disadvantage to its welfare (De Pomerai 1997).

In considering the issue of altering an animal's telos, the UK Banner Committee concluded in 1995 that it would be unacceptable to use genetic modification to increase the efficiency of food conversion in pigs by reducing their sentience and responsiveness, thereby decreasing their level of activity. The committee agreed that it would be morally objectionable to treat animals only as raw materials upon which humanity's ends and purposes could be imposed regardless of the ends and purposes which are natural to them.

In Sweden legislation was introduced which stipulated that farm animals must be allowed to live their lives in accordance with their telos. That cattle have the right to graze and that chicken and pigs have the right to freedom of motion (Straughan 1999: 21). Under Dutch law, telos is also an important issue in deciding whether or not transgenic animal research should be allowed (Brom and Schroten 1993; De Pomerai 1997).

In summary, when considering the creation of different human-nonhuman combinations and the modification of an animal's telos, it may be appropriate to grade these combinations on several scales of increasing severity. In other words:

- A scale of the species involved, allowing more radical manipulations only on less sentient species such as nematodes or fruit-flies.

- A scale involving welfare and telos considerations which is balanced against the anticipated benefits that might accrue if the research went ahead.

- A scale involving the nature of the proposed manipulation. For example, does it only involve the genes already present in the animal genome, or does it involve novel genetic material from a closely related or distantly related species?

The precautionary principle

The precautionary principle is a method of dealing with uncertainty when possible risks are being considered. It states that if a new action or policy may cause severe or irreversible harm to an individual, a community, the general public or the environment, in the absence of full scientific certainty that harm would ensue, the burden of proof falls on those who would advocate taking the action.

The principle exists in several different forms. One of these was adopted by the European Commission of the EU in its Communication on the precautionary principle which was published in 2000. This indicated that:

the precautionary principle forms part of a structured approach to the analysis of risk, as well as being relevant to risk management. It covers cases where scientific evidence is insufficient, inconclusive or uncertain and preliminary scientific evaluation indicates that there are reasonable grounds for concern that the potentially dangerous effects on the environment, human, animal or plant health may be inconsistent with the high level of protection chosen by the EU. (DGHC 2000; Commission of the European Communities 2000)

Another version of this principle was set-out in Principle 15 of the 1992 Rio Declaration on Environment and Development, which states that:

Where there are threats of serious or irreversible damage, lack of full scientific certainty shall not be used as a reason for postponing cost-effective measures to prevent environmental degradation.

The precautionary principle, while subject to an intense debate (HCSTC 2006; Nuffield Council on Bioethics 2003: 57–9), is nevertheless often applied to biological procedures because changes cannot be easily contained and may affect everyone. In the case of technological innovation it may be all the more difficult to contain the impact because of the possibility that the technology can self-replicate.

Application of the principle modifies the status of innovation and risk assessment. It is not the risk that must be avoided or amended, but a potential risk that must be prevented. Legal ethicist Roberto Andorno indicates that:

If there are good reasons, based on empirical evidence or plausible causal hypothesis, to believe that damage might occur, and given the crucial importance of what is at stake ... adequate measure should be taken as soon as possible to prevent such disastrous outcomes. (Andorno 2004: 12)

However, it should be remembered that no activity or process can ever be guaranteed to present absolutely no risk and to be completely safe.

With the precautionary principle, the onus of proof rests on those creating the hazards who should demonstrate that they have appropriately examined the nature and extent of any potential risks arising from their proposed products or activities and that these present an acceptable level of safety (Andorno 2004: 19). Some commentators suggest that there are two forms of the principle, which they call the 'strict form' and the 'active form'. The former requires inaction when a new action might pose a risk, while the latter means choosing less risky alternatives when they are available, and taking responsibility for potential risks.

Environmental concerns

The precautionary principle has been developed primarily in relation to environmental concerns about new technologies. Some of the harm that has been done to the environment by industrial and technological innovations is irreversible. This is most obviously the case when pollution or overuse of resources leads to the extinction of species and the reduction in biodiversity.

In addition to the danger of releasing chemical pollutants or radioactive material into the environment, the introduction of novel species (animals, plants or micro-organisms) into an ecosystem can also be a form of pollution, causing dramatic and sometimes catastrophic effects. This has been documented many times in history as a deliberate or accidental consequence of the movement of human populations.

It is possible to consider the moral relevance of interspecies experiments not primarily in relation to the welfare of the animals used in the experiments but in relation to the effect the experiments might have on 'wild' species. Whereas welfare concerns focus on higher animals, environmental concerns focus on any living forms, and particularly on simpler animals, plants or micro-organisms that can proliferate rapidly in the wild. From this perspective, interspecies experiments should be regarded as a 'biohazard' which needs to be justified by a strong rationale and needs to be contained so that the risks of environmental contamination is minimized.

The proportionality principle

In order to find a resolution between different claims, the proportionality principle may be considered, which examines the ends, means and effects of a specific proposal in a reasonable, consistent and transparent manner. This takes place through a careful assessment of the characteristics of a claim and usually takes into account the three following features (Nuffield Council on Bioethics 2007: 34):

A balancing test which evaluates the aims that are being suggested against the means used to achieve these aims. For instance, the scientific benefits or advantages expected from a procedure being considered should be weighed against the perceived resulting disadvantages or risks.

A necessity test which indicates that if an aim can be obtained through more than one means, then the means that causes the minimum amount of harm should be chosen. This may be considered from a number of different settings such as the effects on society or a specific community.

A suitability test which considers whether the means used are appropriate to achieve the proposed aim.

According to the proportional principle, an act can be justified if the overall good involved in doing the action compares favourably with the overall disadvantages which it would bring about. Alternatively, an action is not undertaken if the overall disadvantages compare unfavourably to the overall benefits which it is considered to bring about. The amount of verifiable information coming in support of a proposed aim will also be crucial in determining whether the suggested aim is proportionate to the obtained results.

In short, the proportionality principle is a tool assisting in moral decision making according to which an agent ought to choose – through a preliminary assessment – that alternative course of action which promises the greater proportion of good over disadvantages (Lawler, Boyle and May 1985: 78–97).

The spectre of eugenics

It was noted in the historical section that early experiments seeking to create human-nonhuman combinations had an explicitly racist and eugenicist background. Those chosen for these early experiments were from 'inferior' races that were alleged to be closer to apes. Similarly, the use of the term 'monster' by Glanville Williams explicitly identified disabled human beings as belonging to a subhuman category, a category originally based on the mistaken belief that severely disabled infants were the result of human-nonhuman sexual union.

There is no necessary link between approving of human-nonhuman combinations and eugenics. Adolf Hitler, who was an obsessive eugenicist, was opposed to human-nonhuman combinations. Likewise it is perfectly possible for a scientist to wish to experiment on human-nonhuman hybrid embryos even though he or she is personally opposed to eugenics. Nevertheless, the cultural connection between human-nonhuman combinations and eugenics is not only a phenomenon of the early twentieth century history. The connection has been perpetuated because interspecies somatic cell nuclear transfer involves the same technology as reproductive cloning, and the creation of transgenic embryos involves the same technology as germ-line genetic engineering in human beings. If someone is concerned that society may be moving towards eugenics without fully realizing it, then he or she may see the technology of human-nonhuman combinations as a step in that direction. Indeed it is arguable that the creation of transgenic human embryos may not only be seen as a case of dysgenetics (the genetic deterioration or weakening of a population) but as an example of eugenics.

The ethical evaluation of eugenics is a significant undertaking for which it is necessary to engage with serious philosophical defences of eugenics as well as philosophical and political critiques. The aim of this section is not to establish a case against eugenics but only to point out that, if someone has ethical or political concerns about eugenics, then he or she will have reason to be concerned about at least some proposals in the area of human-nonhuman combinations.

Human cloning

The immediate context for the proposal in 2007 in the UK to create human-admixed embryos (specifically cybrids) was the cloning of Dolly and the attempt to apply the same SCNT technique to human embryonic stem cell research. The human-admixed embryo debate closely echoed the controversy over cloning human embryos for research.

Cybrids may be regarded as a kind of cloned human embryo, in which case this research represents a step towards cloning a human adult. It is noteworthy that advocates of cybrid research in the United Kingdom cited with approval the work of Zavos and his team (Illmense, Levanduski and Zavos 2006 cited by HFEA 2007b 1.2.11, 4.3.1.1, 4.3.2.14, and 4.4.1.1). However, in 2009 very shortly after the new law passed into effect, Zavos claimed that he had made progress towards the goal of reproductive cloning and that he was experimenting with cybrids in order to improve the cloning technique. This seemed to vindicate those who had warned that cybrid technology, if allowed to develop, would lead to reproductive cloning. While the government maintained its opposition to reproductive cloning, it is also noteworthy that the Human Fertilisation and Embryology Act 2008 repeals the Human Reproductive Cloning Act 2001 and contains a loophole that could legally permit reproductive cloning: the exception was deliberately introduced to allow cloning-like procedures if these prevent the inheritance of mitochondrial disease.

This concern takes the form of a version of the 'slippery slope argument'. It is undeniably the case that the technology of SCNT is necessary for human reproductive cloning and that the research on SCNT de facto supports reproductive cloning. The attempt to achieve SCNT with nonhuman ova both presupposes and promotes work on SCNT more generally. What makes this argument of more than theoretical significance is the fact that one of the very few active researchers in human-nonhuman cybrids (Illmensee) is a collaborator with one of the few scientists actively seeking to perfect human reproductive cloning (Zavos).

It is not within the scope of this book to examine the ethical implications and possible consequences of human reproductive cloning. Others have examined this in great depth (see President's Council on Bioethics 2002). The relevance of this topic to the present work is simply that one form of human-nonhuman combination (cybrids) relies on the same technology (SCNT) as cloning and therefore the facilitation of this research will bring closer the possibility of human reproductive cloning. This is a realistic concern in relation to cybrids, nevertheless it should be acknowledged that this issue is not directly relevant to other forms of human-nonhuman combinations such as chimeras or 'true hybrids'.

The concept of 'ordre public'

Though there is no universally accepted notion of 'ordre public' the concept is useful to consider since it includes the protection of human, nonhuman and plant life in addition to health and may be applied to the subject matter that may lead to serious prejudice to the environment.

This concept is already established on the international stage with, for example, the European Patent Office Guidelines for Examination, Part C, Chapter II, 7. Prohibited matter, 7.2 Matter contrary to ordre public or morality which state that a patent cannot be accepted if a subject matter may undermine 'ordre public'. Examples of the kind of subject matter coming within this category are:

- Incitement to riot or to acts of disorder;
- Incitement to criminal acts;
- Racial, religious or similar discriminatory propaganda; and
- Grossly obscene matter (EPO 2010, C.II-14, 7.2).

Ordre public was the reason given for excluding from patentability processes for cloning human beings and processes for modifying the germline genetic identity of human beings. Interestingly, it was also the reason for excluding from patents the production of human-nonhuman combinations. Directive 98/44/EC of the European Parliament and of the Council of 6 July 1998 on the legal protection of biotechnological inventions indicates in preamble (38) that:

Whereas the operative part of this Directive should also include an illustrative list of inventions excluded from patentability so as to provide national courts and patent offices with a general guide to interpreting the reference to ordre public and morality; whereas this list obviously cannot presume to be exhaustive; whereas processes, the use of which offend against human dignity, such as processes to produce chimeras from germ cells or totipotent cells of humans and animals, are obviously also excluded from patentability...

Traditionally, 'ordre public' in the USA legislation was referred to the subject matter that was 'frivolous or injurious to the well-being, good policy, or sound morals of a society' (Lowell v. Lewis 1817). This raises the question of what constitutes 'sound morals of a society' and, in particular, whether the creation of human-nonhuman entities is intrinsically, in itself, a transgression of a sound moral principle.

Intrinsic concerns

Intrinsic concerns are generally considered as being more basic and funda-mental than extrinsic ones. If a practice is considered to be intrinsically wrong, this determines the basic question of rightness or wrongness of the practice without the need to consider extrinsic pros and cons. Indeed, further considerations of consequences or intentions become morally relevant only as additional factors (Straughan 1999: 19). In this regard, the concepts of unnaturalness and human dignity will be examined in the following sections.

Unnaturalness

Some opposition to the crossing of the species barrier is based on the belief that it would contravene the order of nature, in which every species is distinct with an essential and unchanging character. This type of thinking can be seen in Aristotle's (384–322) idea, also developed by St Thomas Aquinas (c.1225–1274), that all living things tend towards their own inner ends or goals (telos), and their biological functions exist to help them achieve these goals. From this perspective, the morally correct action is to ensure that each species is aligned with its respective goals (Kass 1985: 249–75; Heeger and Brom 2001). As merging human and nonhuman tissues at the embryonic stage would hinder both beings from fulfilling their fundamental purpose, it would therefore be seen as unnatural and wrong (Karpowicz, Cohen and Van der Kooy 2004).

There is also an argument which states that, while it is not always clear, there must be a good reason why the distinction between species exists in nature, and that interfering with this order could have serious – but as yet unknown – consequences on both a biological and a social level. To cross the species barrier could also be opposed from the perspective of aesthetics or from arguments of species integrity.

One problem with this argument is that biologists disagree about the extent and even the existence of 'natural species boundaries'. Indeed, the meaning of the word 'species' is itself unclear, and depends on the context in which it is used. Natural crossing of species boundaries already occurs to an extent without any help from modern biotechnology. According to estimates, about 10 per cent of wild bird species cross fairly frequently and in some cases higher rates of over 30 per cent are observed, for example in the birds of paradise and among Darwin's finches (Straughan 1999: 12).

It is sometimes alleged that concern that human–nonhuman combinations are 'unnatural' or transgress an 'order of nature' represents an undue attachment to pre-Darwinian categories and is a failure to acknowledge that human beings are animals and products of evolution. However, it is feasible that the opposite may also be taking place. Crossing the species boundary is biologically possible precisely because of underlying relationships between different species, and this kind of cross is of moral concern precisely because human beings are animals of a particular species. The animality of human beings is easily forgotten. Even the Human Fertilisation and Embryology Act 2008 contains the stipulation that 'For the purposes of this section "animal" is an animal other than man' (Human Fertilisation and Embryology Act 2008, 4A (8)), so it is denied that humans are legally 'animals'! However, it is precisely because human beings are animals of a particular kind that they have a particular character and flourish in particular ways.

It is important to consider here what is meant by 'unnaturalness'. Nature, by itself, does not come with a set of clear ethical guidelines that can be consulted relating to the typical functioning and purpose of a living being (Karpowicz, Cohen and Van der Kooy 2005: 114). If one means 'artificial' or 'man-made', then almost every aspect of Western civilization would be 'unnatural', and more traditional products and processes would be deemed just as 'unnatural' as the mixing of species (Straughan 1999: 12). Similarly, if the term refers to anything which does not occur in nature, then taking medication for an illness, which most people see as perfectly acceptable, would also be viewed as unnatural because the human body does not 'naturally' heal itself that way (BACS 2008). It is possible to argue that nothing people do, including scientific experiments, is any more or less unnatural than anything else.

It is also possible, however, to make a distinction between facilitating or augmenting a natural process and manipulating nature to do something that seems to undermine or contradict the natural gaols of health, integrity and flourishing. It is because of their biological nature that people wish to be parents and go to great lengths to conceive offspring of their own. It is for similar reasons that children wish to know about their biological origins and family history. Notwithstanding the hugely important role played by adoptive parents, the fertility industry has grown up to answer a desire of couples and of individuals to beget children of their own. Human beings are animals with a particular nature and human cultural and intellectual achievements build upon this nature. Cloning and the creation of interspecies hybrids arguably fall into the second category. What is clear is that many people are concerned at the perceived lack of any respect for natural limits such as species barriers in the pursuit of scientific advancement. The reasons for public concern are, moreover, not limited to conventional and academic ethics. Although philosophers often dismiss the 'naturalistic fallacy', in ethics, the question of naturalness remains relevant in considering the wisdom of particular scientific interventions.

In relation to creating newborn creatures, where there is a significant proportion of human and nonhuman characteristics there seems to be an instinctive and intense revulsion. This is as evident in defenders of cybrid research as it is among opponents. Virtually no one defends the freedom of a scientist to seek to bring such a creature to birth. Such emotional reactions against creating a monster do not settle this issue, but they do provide a starting point. Emotional reactions may be appropriate, or they may be inappropriate, and to decide which '[w]e must spell out the message of the emotions and see what they are trying to tell us' (Midgely 2000: 9).

In a letter to *Nature* in 2003 on the subject of reproductive cloning, one correspondent commented that, 'I personally am completely opposed to human cloning of any sort, from a feeling of utter repugnance towards what appears to me to be a fundamental assault on human dignity, with the potential for horrendous misuse' (Leader 2003). This comment is helpful because it moves beyond repugnance to the category of human dignity, a concept frequently invoked in bioethics. The same move was made in 1998 in relation to the patentability of human-nonhuman combinations: 'processes, the use of which offend against human dignity, such as processes to produce chimeras from germ cells or totipotent cells of humans and animals, are obviously also excluded from patentability' (Directive 98/44/EC of the European Parliament, paragraph 38). This argument was also invoked by Karpowicz and colleagues in 2005:

> By giving nonhumans some of the physical components necessary for development of the capacities associated with human dignity, and encasing these components in a nonhuman body where they would either not be able to function at all or function only to a highly diminished degree, those who would create human-nonhuman chimeras would denigrate human dignity. (Karpowicz, Cohen and Van der Kooy 2005: 121)

It was reiterated by some critics of the United Kingdom's legalization of human-nonhuman embryo research. 'From the ethical standpoint such procedures represent an offense against the dignity of human beings on account of the admixture of human and animal genetic elements capable of disrupting the specific identity of man' (CDF 2008, paragraph 29). What follows is an evaluation of the claim that the creation of human-nonhuman admixtures contravenes the full inherent dignity of the human being (see also Jones 2010).

Full inherent dignity

Like many other terms in ethics and philosophy, 'dignity' has often been used as an empty slogan, or a cover for intellectual shortcomings. Commenting on the appearance of this vague usage in connection with end of life treatment, a US presidential commission observed: 'Phrases like ... `death with dignity' ... have been used in such conflicting ways that their meanings, if they ever were clear, have become hopelessly blurred' (President's Commission 1983: 24). It is true that the concept of dignity cannot simply be equated with other concepts such as respect and autonomy, beneficence, non-maleficence or justice. But this does not invalidate the basic idea.

The *Oxford English Reference Dictionary* defines dignity as the 'state of being worthy of honour and respect' (Pearsall and Trumble 1996). In other words, it incorporates aspects of 'honour' and 'respect' but also of 'value' and 'worth'.

The idea that human beings possess a special status among animals is certainly ancient, being implicitly and explicitly affirmed in both secular and religious writings and practices. In more recent philosophy, however, a development of the term 'dignity', as referring to the fundamental moral quality of human beings, can be traced back to Immanuel Kant's (1724–1804) *Groundwork of the Metaphysics of Morals*, published in 1785.

In seeking to define the foundations of the human moral sense, Kant proposed that there is essentially a central moral obligation (termed the Categorical Imperative) which everyone should fulfil and from which all others can be derived. Accordingly, for Kant, one ought to act 'only according to that maxim through which you can at the same time will that it should become a universal law' (Kant 1964: 52/421).[2] From this, Kant derived a second formulation of the Categorical Imperative whereby one should: 'Act in such a way

---

[2] References to page numbers for Kant's *Groundwork* are given here are first to the second German edition (the best published in Kant's lifetime) and second to the edition of the Royal Prussian Academy in Berlin, in this case page 52 and page 421 respectively.

that you always treat humanity, whether in your own person or in the person of any other, never simply as a means, but always at the same time as an end' (Kant 1964: 66/429).

This central obligation applies to all rational agents, which for Kant was equal to all human beings. In other words, rational autonomous agents generate moral obligations towards each other. From this perspective, rational autonomy (the capacity for free obedience to the moral law of which one is the author) is also the basis of the 'dignity of human nature and of every rational nature' (Kant 1964: 79/436) 'dignity' being the partner concept to the Categorical Imperative. Thus, for Kant if a rational and autonomous being has 'an absolute worth' and is 'an end in itself', these qualities give the being dignity, which is an immeasurable quality, and makes him or her moral as well as capable of preparing laws. In other words, human beings have dignity, because of autonomous reason rather than impulses or the pursuit of personal or social benefit.

It is the important association which Kant draws between rational autonomy and moral obligations that justifies the moral differences between humans and nonhuman animals, as nonhuman animals are not, for Kant, rational agents. However, he would attribute dignity to any autonomous rational being and not specifically to a human person.

The following passage demonstrates Kant's link between human dignity and morality itself.

> Suppose, however, there were something *whose existence* has *in itself* an absolute value, something which as *an end in itself* could be a ground of determinate laws; then in it, and in it alone, would there be the ground of a possible categorical imperative – that is, of a practical law. (Kant 1964: 64/428)

The extent of Kant's influence on western philosophy is such that one can suspect that, were it not for his concept of human dignity as the basis of moral obligations and responsibilities, the concept would not be expressed in the same way in so much international legislation.

An emphasis on human dignity in many post Second World War legal instruments also arose in order to address the abuses that took place during this war. After the racism and genocide of the Nazi regime, the international community firmly declared the inherent equality and value of all human beings. Human life was celebrated as the ultimate valuable, its protection was raised once more to the top of the list of state concerns, and it was the concept of human dignity that grounded this.

As a result, the concept of human dignity was endorsed in a number of post-war legislative documents including the Universal Declaration of

Human Rights (1948), the German constitution (1949) and, more recently, the Universal Declaration on Bioethics and Humans Rights (2005). These documents seek to set down principles to govern the manner in which human persons treat each other and more specifically in the context of judgements about their value. Human dignity is seen as a foundation for human rights, and for an egalitarianism that rules out exploitation or persecution of one group by another. In the various declarations and conventions that mention human dignity, though they invariably separate it from human rights and/or the inviolability of human life, they almost always mention them together. In this manner, it has been suggested that human dignity is a foundational concept on which other ethical pillars rest.

The concept of human dignity, however, can be defined in different but complementary ways. In other words, it is possible to characterize different types of human dignity (Sulmasy 2007) which can be separated into two broad categories. The first, 'inherent dignity', reflects an unconditional, inviolable, inalienable and indivisible worth and quality that belongs equally to all members of humanity. It is a permanent feature or characteristic that is not acquired.[3] This sense of dignity is in line with the UN Universal Declaration of Human Rights which affirms in its preamble 'the inherent dignity and ... the equal and inalienable rights of all members of the human family' as 'the foundation of freedom, justice and peace in the world'.

In this respect, it should be noted that the colloquial use of human dignity is often (intentionally) vague since more precise definitions might risk dividing public discourse and damaging consensus where unity is desirable.

Human dignity, moreover, is a concept that cannot be really defined or completely evaluated (Andorno 2001). The most rigorous terminology can acknowledge but not explain the concept. It is something that cannot be reduced to scientific enquiry and is simply accepted as being shared by all people. In this context, human dignity is related to a given reality, intrinsic and unique to the human substance that entitles every human being to a higher moral status than the rest of the natural world. It is not dependent upon any varying functional capacities such as intelligence, abstract reason, language, creativity, ability to feel pain, empathy, awareness of personal biography over time, health or beauty and quality of life but should reflect the dignity that the person already holds.

In contrast, the second type of dignity is 'non-inherent', meaning it is a contingent, variable condition upon which a person merits various honours or a certain amount of respect in different contexts. Non-inherent dignity may

---

[3] The permanent feature of the 'inherent' concept may be contrasted, here, to the 'intrinsic' concept which reflects that something belongs, instead, 'from the inside' though both notions are very similar.

be gained or lost depending on the comportment of an individual and is often characterized as 'personal dignity', but inherent dignity is permanent.

In the rest of this chapter, it will be the concept of inherent human dignity that will be examined since it is this type of dignity that is intrinsically related to basic human rights. Indeed, in the study of the possible moral status of interspecies entities, it is the inherent nature of this dignity that is, by far, the most consequential in ethical discourse though non-inherent dignity may still be important.

In the context of human-nonhuman entities, moreover, talking about a specific inherent human dignity, as such, may not be appropriate since many of these entities are only partly human and to different degrees. In other words, it may be preferable to just use the term of full inherent dignity in the subsequent chapters, which is associated with having a full moral status. In this way, human dignity would only be one 'kind' of full inherent dignity which is associated with the species *Homo sapiens*. Some human-nonhuman entities may then have a dignity which is not 'human' but still have the same full moral worth as inherent human dignity in its inviolable and immeasurable qualities while being completely different to that of nonhuman animals.

This may also go in the direction of partially addressing any accusations of 'human racism' or 'speciesism', made by some commentators (Ryder 1971; Singer 1976), whereby only humans are ascribed this full inherent dignity. This is because, other beings who are not completely (or not at all) members of the *Homo sapiens* species may still be recognized as having full inherent dignity in certain cases.

But a being cannot have a degree of this full inherent dignity. There is no intermediate option. For instance, even though certain nonhuman primates are increasingly being protected in national and international regulations, such as in the UK and the Council of Europe, they are still not recognized by society as having this full inherent dignity. Instead, they have an inherent dignity, but one which is of a different nature to the full inherent dignity of human beings.

Personhood

In the ancient world, the concept of personhood was not something that was recognized or seen as relevant. For example, in ancient Greco-Roman philosophy a clear concept of personhood was completely missing. In Greek Platonic thought, the 'individual' is secondary to the reality of ultimate forms. Humankind's immortal soul is what is primary; a time of attachment to a body is secondary (Zizioulas 1991: 36). In Aristotle's schema, though there is a clear emphasis on the individual, with the soul being the form, 'the organising principle' of the body, there is no permanence of the individual beyond death.

The Greek term for person (*prosopon*) first appeared in the context of theatre, where it referred to the actor's mask. Although it soon also referred to the actor's role and identity as the age-old topics of humankind's existence, meaning and purpose were played out. But an individual's essential being was never identified with his role (person). Similarly, in Roman thought, persona referred to the role one played, rather than any deeper existential meaning.

These views on the human person were also reflected by Greco-Roman cosmologies. What dominated was an over-arching holistic unity where everything formed an inherent oneness, including God, the world and all individuals, which was defined as the cosmos. It was only when Judeo-Christian scholars argued that the Creator and creation were radically distinct that philosophy was eventually liberated from such an all-inclusive cosmology.

The earliest philosophical definition of personhood came from the Roman philosopher Boethius (c. 480–524), who defined a person as an 'individual substance of a rational nature' (Boethius 1918: 85). This definition predominated for more than a millennium and allowed Christian scholars to define various types of persons, such as the three persons of the Godhead, who do not always correspond to the biological category *Homo sapiens* (Eberl, 2007).

At the end of the seventeenth century, the English philosopher John Locke offered an alternative definition of a person as 'a thinking intelligent being, that has reason and reflection, and can consider itself as itself, the same thinking thing in different times and places' (Locke II.XXVII.9) which also allows for the possibility of nonhuman persons.

In this regard, it can be suggested that there is a very close relationship between personhood and full inherent dignity with the possibility of persons being defined as beings to which full inherent dignity is conferred.

Full inherent dignity in the context of rationality and autonomy

The Roman politician and philosopher Marcus Aurelius (121–180) and even earlier Stoic philosophers believed that human beings have basic equality embedded in their common ability to reason. A classic statement of a form of this outlook is found in Blaise Pascal's (1623–62) summary observation that 'all our dignity consists then in thought' (Pascal 1910, fragment 347; Marcus Aurelius 2003: 167).

As noted earlier, the concept of rationality and autonomy as the basis for full inherent dignity is especially portrayed in the philosophy of Immanuel Kant who argued that the fundamental moral motive – the Categorical Imperative – arose from rationality and autonomy. Thus, according to Kant, each person is worthy of respect because he or she is as a self-legislator giving the moral law unto himself or herself.

However, Kant's explanation of full inherent dignity remains at a purely formal level and does not give a satisfactory answer as to the origin and foundation of this dignity (Andorno 2001). Moreover, it is far from clear why self-legislation entails full inherent dignity in any ordinary understanding of the word and not other characteristics such as generosity and a capacity to love. In other words, if the Kantian reverence for dignity only expresses reverence for autonomy, how can the principle of dignity bring additional reasons for assisting the weak, into the field of bioethics and law? (Kemp 1998: 8) The interpretation of Kantian texts is not unanimous. For other specialists, even adopting a Kantian position one cannot deduce from the principle of autonomy that the weakest human beings are not persons (Beckmann 1998: 146). For example, are some human beings who are not rational beings, such as newborn infants, the profoundly mentally disabled and the hopelessly comatose, not endowed with inherent dignity? Is this conclusion not contrary to common intuition? Is it not a paradox that 'moral autonomy', which is one of the most relevant signs of inherent dignity, could be used against it? (Andorno 2001; Jones 2009c). If only this manner of defining full inherent dignity is used it could come into conflict with the Universal Declaration of Human Rights, whereby '[a]ll human beings are born free and equal in dignity and rights'.

Some legal philosophers, such as Deryck Beyleveld and Roger Brownsword, have sought to develop Kant's proposals concerning dignity by suggesting a reason-based approach to inherent dignity (Beyleveld and Brownsword 2001). In doing so, they continue to support Kant's proposal to embed full inherent dignity in an individual's reason and capacity to be moral agents, but then adapt the moral philosophy of Alan Gewirth by emphasizing that agency is actually based on choice (Gewirth 1978). In other words, they indicate that 'the essence of the dignity of agents resides in their capacity to choose, to set their own ends' (Beyleveld and Brownsword 2001, p. 5). Accordingly, these philosophers consider dignity more as an empowerment than as a constraint since the concept of reason is the basis for protecting each individual's right to choose and have autonomy. This is significantly different, however, from a Kantian emphasis on people being 'ends in themselves' which places an important protective constraint and limit on how people may be treated because of who they are (Mitchell et al. 2007: 64–5).

As an example of the clash between these two concepts of dignity, Beyleveld and Brownsword use a dwarf-throwing contest in France, in which the police were eventually authorized to stop the attraction in clubs even though all the individuals taking part in the event had given their consent (Meilander 2007: 33–52). Thus, Kantian dignity may be used to control (or prohibit) activities to which an individual freely consents and which seems to harm no one else.

Personhood in the context of rationality and autonomy

A number of philosophers have also proposed that a person is any being that exhibits the capacity for self-conscious rational thought, augmented perhaps by other capacities such as using language to communicate, having non-momentary self-interests and possessing moral agency or autonomy (Tooley 1983: 146; Singer 1992: 84; Warren 1994: 308). This view is sometimes described as the 'performance theory'.

As Kant noted:

Beings whose existence depends, not on our will, but on nature, have none the less, if they are non-rational beings, only a relative value as means and are consequently called 'things' [*Sachen*]. Rational beings, on the other hand, are called 'persons' because their nature already marks them out as ends in themselves – that is, as something which ought not to be used merely as means – and consequently imposes to that extent a limit on all arbitrary treatment of them (and is an object of reverence). (Kant 1964: 65/428)

According to this definition, personhood is independent of membership in the species *Homo sapiens* (Singer 2005: 41). In other words, not only might some nonhumans be considered persons, but some humans may also not be considered persons because they do not exemplify the relevant capacities. Examples include foetuses, anencephalics and irreversibly comatose patients. Insofar as performance theorists define the boundaries of a person's existence according to the exemplification of self-conscious rationality, the criterion for determining such boundaries is, at minimum, the presence of a fully formed and functioning cerebrum (Eberl 2007).

Toward the end of the seventeenth century in his 'Essay Concerning Human Understanding' the philosopher John Locke wrote:

We must consider what person stands for; which I think is a thinking intelligent being, that has reason and reflection, and can consider itself the same thinking thing, in different times and places; which it does only by that consciousness which is inseparable from thinking and seems to me essential to it; it being impossible for anyone to perceive without perceiving that he does perceive. (Locke II.XXVII.9)

Another view is reflected by the medieval philosopher Thomas Aquinas (c. 1225–74), who adopts Boethius's definition of personhood, which concurs with performance theory insofar as 'rationality' is taken as its definitive feature (*Summa theologiae* Ia, Q. 29, a. 1). However, Aquinas differs from the performance theorists because he holds that every human being is a person (*Summa theologiae* IIIa, Q. 16, a. 12 ad 1). Some commentators have also interpreted Aquinas' views by suggesting that one should consider all the relevant capacities for the definitive rational activities of persons at every stage of his or her existence, whether before or after death. At the embryonic stage, these capacities are present because the embryo has the complete human genome and other intrinsic biological factors that are necessary and sufficient – given the right supportive uterine environment – for it to develop into an actually self-conscious rational being (Eberl 2005, Eberl 2006: 23–42).

Combining these disparate views, one can propose the essential properties of personhood as either the capacity to actually exhibit self-conscious rational thought (performance theory) or the capacity to exhibit self-conscious rational thought given time and proper development both before and/or after death.

Full inherent dignity and relationships

Another important and complementary understanding of inherent dignity can be considered in the context of relationships. In other words, human persons have dignity because they are given that dignity in the context of a relationship with at least one other person. The investment of dignity by a person to another, and humanity in general, through the means of a relationship is one of the most fundamental dignity-giving concepts. From this perspective, beings have dignity as a matter of contingent fact. Human persons treat each other in a certain manner, and expect to be treated in such a manner, and so legislation should represent that reality.

Generally, a psychologically healthy and conscious person gives himself or herself full inherent dignity. Thus, dignity is primarily conferred through the relationship a self-aware person has with himself or herself. The importance for a person to always maintain his or her own inherent dignity can also be consistent with the emphasis on duty found in Kantian ethics.

Many religious believers base their notion of full inherent dignity on their belief that human beings are created by, and have a special relationship with, a transcendent being. For example, the Biblical creation story places humankind in a different stage of creation from that of all other life, and indicates a special relationship that God has for human persons who are created 'in his own image', that animals do not have. Thus, human persons are distinct from animals in having a special dignity.

This proposition, or a similar one, is held by all the major monotheistic faiths and has been seen as one of the premises of the religious affirmation of human dignity. (For example, Muslims believe this takes place through the 'caliphate' in which the human person is considered to be a representative of God on earth.) In this case, it is because a relationship can exist between this transcendent being and a human being, such as an infant or the severally mentally disabled individual that human dignity is conferred to this individual even though this relationship may only be in one direction.

Personhood in the context of relationships

For many commentators, personhood can only exist if a relationship is present between beings (where at least one of these beings is rational). In this case, it is not only the existence of the being that matters, but also the relationship(s) that are made possible because of this existence. It is only from this perspective of relationships that human beings who are not rational, such as foetuses, infants, the profoundly mentally disabled and the hopelessly comatose, are endowed with full inherent dignity.

But this relationship may also be self-reflective. In this regard, the ethicist John Harris notes that most of the characteristics for personhood suggested by Locke such as intelligence, self-consciousness, memory and being able to conceive of the future can all be considered as being variable and existing to different degrees in different individuals. Thus because he rejects the concept that personhood can be variable he suggests instead that a person can only be 'a creature capable of valuing its own existence' (Harris 1999). In other words, a self-conscious person who is able to give himself or herself inherent dignity in a sort of self-reflective relationship. However, this still means that embryos, foetuses and infants but also adults affected by serious mental disorders as well as individuals who do not value their own lives and who may be suffering from suicidal depression would not be considered as persons (Cobbe 2011). On the other hand, according to Harris, extra-terrestrials, gods, angels, devils and certain animals or machines, may be included in the definition of a person since they are capable of valuing their own existences.

Mixing species, speciesism and threats to inherent dignity

As the previous discussion on inherent dignity emphasizes, there is no single attribute to what defines this dignity. Instead, it is a multifaceted notion that cannot simply be reduced to capacities. Because of this, the concept is difficult to clearly determine though it remains crucial in the understanding of personhood and is the very basis of civilized society.

In this context, it has been suggested that undermining the biological distinction between human and nonhuman animals threatens the moral distinction between human dignity and the manner in which other animals are considered. In other words, that the production of creatures in which the lines between human and nonhuman are blurred could create moral confusion undermining the very concept of human dignity and the high moral status with which human beings are ascribed (National Academy of Sciences 2005: 55; Robert and Baylis, 2003). Others have indicated that if a nonhuman animal could be created that may exhibit, to some degree, human capacities relevant to human dignity, then such animals should not be created (Karpowicz, Cohen and Van der Kooy 2005; Johnston and Eliot 2003).

Indeed, it seems that the attempt to create a part human, part nonhuman being would be wrong to that being. What would happen, for example, if a cross between a human and a chimpanzee was created, if such a cross were possible? How would the newborn 'humanzee' be regarded? Would it be a child or a nonhuman animal? Should he or she go to a special school? Should he or she be reared in a human environment? Should he or she be clothed? If he or she were subject to experimentation or brought up in a laboratory environment (as might be the case if it were created deliberately by scientists) or 'exhibited', would this be an injustice? If there were a dilemma in which it was possible to save either the humanzee or a human child, but not both, would it ever be right to choose to save the humanzee and let a human child die? What is important here is not simply the existence of practical perplexity – for practical questions can be resolved, but perplexity about fundamental moral attitudes. Does this individual share in common humanity and share in the dignity of human nature?

There are situations in nature where familiar categories break down, resulting in perplexities. Aristotle remarks that there are creatures in the sea about which it would be difficult to say whether they are animals or plants (History of Animals, VIII.1). If a creature were discovered that was not human but shared significant human characteristics, then society would have to cope with this ambiguity. However, it is very different deliberately to create a situation of moral ambiguity. The relevance of what is brought about deliberately is evident from less extreme examples of family and genetic identity. Adoption creates a situation of ambiguity about identity and family. It raises difficulties for parents and children to negotiate, especially during adolescence. Nevertheless, adoption represents a positive attempt to provide a family environment for a child who otherwise would lack one. Adoption is totally different from surrogacy arrangements where it is intended, from the beginning, that the surrogate mother will have no place in rearing her child. Surrogacy arrangements have no legal force in the United Kingdom, and the welfare of children is not served by encouraging them. Society lives with ambiguity, sometimes very fruitfully, but this does not justify a deliberate licence to create such ambiguous and potentially damaging situations. In the case of the humanzee he or she would be placed in a no man's land between human and nonhuman.

Would creating a humanzee be making a monster? The word monster comes from something that is shown or displayed (*monstro*), as people with physical deformities have often been displayed in the past. This is a very shameful and recurrent phenomenon in human history. Even the lauded Cambridge academic Glanville Williams could describe a disabled infant as a 'monster' as 'it' rather than 'he' or 'she', as a creature who could 'lawfully be put to a merciful [*sic*] death'. He also described conjoined twins as 'a species of monster' (Williams 1958; Keown and Jones 2008). In the light of the historical use of the term, it would be better to avoid all reference to 'monsters'. Nevertheless, a deliberately created humanzee would surely become a monster in the original sense, an object of human fascination and pity. This would represent a harm to the creature and something that it would be wrong deliberately to bring about.

It seems likely that creating a humanzee would cause problems for the creature and perplexity for society. However, the more fundamental problem seems to be that the attempt to cross the species boundary represents an inhuman act. Just as bestiality is inhuman and a travesty of human sexual union, so deliberately creating a half-human, half-nonhuman creature is a travesty of human procreation. In the past there have been societies that unjustly and unreasonably prevented marriage between people of different ethnic or racial groups. More offensive yet, this injustice was itself advocated on the analogy of forbidding bestiality. As argued above, the idea of human dignity is based on a sense of solidarity, relationships and sharing a common nature (with its common dignity and common misery). This involves respect for all human beings and for the manner of human procreation.

In the context of discussion of reproductive cloning, advocates have pointed out that this 'respect for human procreation' argument was used against *in vitro* fertilization, which is now widely accepted. Nevertheless, while it is arguable that *in vitro* fertilization enables couples to become parents (of their own genetic offspring) and so to participate in one aspect of human flourishing, the same cannot be said of reproductive cloning, and still less of the crossing of the species boundary. These processes do not assist human procreation but constitute a radically different kind of generation. It is for this reason that Tonti-Filippini characterizes such processes as 'an offense against the sacredness of the generative faculty that subsists in the human genome' (Tonti-Filippini et al. 2006: 703). In neither case is the process a sexual reproduction by human parents. In the case of human-nonhuman combinations it is not even clear that the process generates a being with full inherent dignity.

In this regard, most individuals believe that nonhuman animals should not be considered to have a similar dignity to human beings. If this was not the case, animals could not be killed and used for food or other uses without the killing being considered as murder.

On the other hand, the theory of 'animal rights' is sometimes invoked to condemn the existence of a specific full inherent dignity such as human dignity, which is quite different to that of nonhuman animals, as an unjustified anthropocentrism. This is the so-called 'speciesism' (Ryder 1971; Singer 1976) argument, whose name draws parallels with racism. Thus, some commentators have argued that as humanity shakes off the last remnants of its racist past, humanity must deal similarly with its 'speciesist' present.

On this basis, Peter Singer builds a philosophical edifice which assigns an equal moral worth and consideration to sentient animals, capable of suffering, as to other human beings (Singer 1976). As a result, any human action which exploits animals, or which gives preference to human interests over (equivalent) animal interests, is seen as 'speciesist'. Singer believes that the concept of a special and specific human dignity which is of another 'nature' to that of nonhuman animals does not exist. That there is no inherent and firm moral boundary, for instance, between human beings and great apes. In addition, if nonhuman animals are morally equal to human beings in the natural world, there is no need for a special human dignity since dignity is equally recognized to various degrees in all living creatures. In this way, and concerning a severely retarded human infant, who might not achieve the intelligence of a dog, Peter Singer argues that:

> The only thing that distinguishes the infant from the animal, in the eyes of those who claim it has a "right to life", is that it is, biologically, a member of the species Homo sapiens, whereas chimpanzees, dogs and pigs are not. But to use this difference as the basis for granting a right to life to the infant and not to the other animals is, of course, pure speciesism. It is exactly the kind of arbitrary difference that the most crude and overt kind of racist uses in attempting to justify racial discrimination. (Singer 1976: 18)

Singer does not actually argue that if speciesism was rejected it would be acceptable to choose to kill an infant instead of, for example, a dog, if a choice had to be made. Instead, he argues that the characteristics that normal human beings have, such as rationality as well as a concept of past and future, give normal human beings a greater right to live and to be respected than nonhuman animals.

Other philosophers, such as Mary Midgely, accept the driving force behind this argument but come to less radical conclusions, assigning nonhuman animals significant moral status without proclaiming their equality to human beings (Midgely 1983). Singer's concept of sentiency also presents the real biological problem of distinguishing between sentient animals worthy of moral concern, and non-sentient animals which are not. A question could be asked, for example, whether one should draw the line beneath mammals, beneath warm-blooded vertebrates (to include birds), or beneath all vertebrates? But then what about some of the more advanced invertebrate groups, such as various cephalopods (e.g. octopus, cuttlefish) (De Pomerai 1997)?

It has been suggested that if all animals, including humans, are equal in inherent dignity, no human being has full intrinsic dignity, because the notion of full inherent dignity specifically implies a fundamental distinction between the human and the nonhuman realms (Andorno 2001). In this regard, others suggest that the concept of full inherent human dignity is not opposed to the interests of nonhuman animals because it is inclusive, not exclusive. In other words, human dignity is not based on a negative comparison between the human species and other species (Jones 2010: 105; Jones 2010: 22n. 38)). On the contrary, it is compatible with extending inherent dignity to other species, if this were merited, and positively invites the extension of some level of minimal protection (inherent though not full dignity) to all nonhuman species.

One must remember that the moral difference between human and nonhuman animals has generally been presumed throughout the history of law even with nonhuman primates. It is certainly true that nonhuman animals are afforded far less protection in law than human beings. Of course there are exceptions with, for example, some animals being held sacred in certain countries but it remains true that the attitude of law towards human beings is usually different.

Ethical perspective of different body parts

Many persons believe that the ethical implications of implanting foreign body parts into a human individual are related to the degree of change that this may entail in the human identity of the person receiving them (Pontifical Academy for Life 2001). But not all parts of the human body are generally considered to be equally important in the expression of the identity of the person. Some body parts exclusively perform their specific function such as the heart, which may be considered as just a biological pump. Other parts, instead, add to their functionality a strong and personal symbolic element which inevitably depends on the subjectivity of the individual. And others still, such as the brain and reproductive cells, are often strongly linked with the concept of human identity and dignity of a person.

Therefore, many people would not object to purely functional animal parts being transferred to a human person if it was done on a case by case basis, and depending on the specific relation to the symbolic meaning which they take-on for each person. However, concerns have been expressed with regard to the transfer of nonhuman brain and reproductive cells to human persons. In a study organized in 2010 by the UK Academy of Medical Science and others, a public survey found that the possibility of changing the brain of a nonhuman animal may be outside the boundaries of acceptability. This was especially the case if the nonhuman animal's cognition was changed. In the same study, it was also found that any crossing of the boundary between a human and nonhuman animal's reproductive system was seen as unacceptable (Ipsos MORI 2010: 4).

Such public responses may reflect the specific risks connected with different procedures with respect to human identity (Pontifical Academy for Life 2001; Human-Animal Hybrid Prohibition Act 2007). This is because both the animal brain and reproductive cells are related to what is understood in the concept of a specific organism being:

> 'Whole' in the sense that nothing from within has been taken out of the organism which would result in a loss of its existence. That nothing is missing from an inherent perspective. It also relates to the notion of individuality and 'oneness'.

> 'Complete' in the sense that nothing is missing from an external perspective which would result in a loss of existence. It includes for the living organism, the notions of (1) completeness in the three dimensions of space and (2) completeness throughout the time of its existence.

Moreover, the brain and the reproductive cells both form part of this complete and whole organism because they both originated in the same specific generation event of this organism as an embryo.

Thus, on the one hand, it is because a living human person is considered as only remaining complete and whole if he or she has a living brain that this part of the body is believed to have a special moral value. This is reflected in that a person is generally accepted to have died if his or her brain no longer functions. (The subject of brain death will not be addressed in this book though it should be noted that a vigorous debate is taking place regarding the validity of whole-brain death.) The body, in this case, has become 'incomplete' and has lost its 'wholeness'. However, if for example a kidney is replaced with that of another person or animal, then the person is generally considered to have retained his or her completeness and wholeness.

Similarly, it is the reproductive cells that are responsible for creating another specific 'complete' and 'whole' living organism. This is important with regard to the identity of the procreator or creature. If, for example, the ovaries of a woman were replaced with that of another person or animal, questions could be asked relating to the manner in which the offspring would still be considered as having been procreated from the wholeness and completeness of his or her 'mother'.

The brain and human identity

Probably no other organ is considered as important for human identity and the sense of self as the brain. For a person who is aware, the brain is indeed generally accepted as the source of this awareness of his or her own existence and identity. Without a living and functioning brain, the person would not be considered as being alive or have a biological personality. In addition, if this brain was to be significantly modified so that an interruption took place in the continuity between the past and present sense of 'self' and identity of the person, then the present person could even begin to question whether the past identity had not, in some way, been lost.

For instance, if the transplantation of animal neural cells into a person's brain resulted in this person no longer recognizing their past sense of self and identity, they could maintain that the previous person no longer existed and that a new person, with a new character and sense of self was now present. This would result in some profound ethical but also philosophical problems in the understanding of what constitutes a person and their own identity.

Alternatively, if a nonhuman animal was to obtain a certain kind of brain which made it self-aware, a question would then arise whether this being should not have similar rights to human beings. This would create deep ethical unease and would make a re-evaluation of what makes human beings entitled to full inherent dignity, such as human dignity, necessary.

In this regard, an interdisciplinary working group in the USA published, in 2005, the results of its consultation on the scientific, ethical, and policy issues raised by research involving the engraftment of human neural stem cells into the brains of nonhuman primates (Greene et al. 2005). The group indicated that the introduction of human cells into nonhuman primate brains raised questions about the very nature of the moral status of a being. They also acknowledged that, to the extent that a nonhuman primate attains certain capacities, it may be argued that such a creature should be held in correspondingly high moral standing. However, the group did not clarify whether it believed that such primates might actually become persons or whether, remaining nonhuman (and non-personal) primates, they should simply be accorded special treatment (Berg 2006).

In this context, a variety of reasons are often given for according different moral standings to human and nonhuman primates depending on different worldviews. The philosopher John Stuart Mill (1806–73), considered the richness of human mental life to be an especially fecund source of utility (Mill 1987: 83). Peter Singer, although strongly defending equal consideration of nonhuman interests, argued that self-awareness affects the ethically allowable treatment of a creature by changing the kinds of interests it can have (Singer 1979). Thus, many of the widely accepted candidates for determining moral status, in the secular world, involve mental capacities. These include (Shreeve 2005):

- The ability to experience physical pleasure and pain as well as psychological joy and suffering,

- The use of language,

- Rationality,

- The possibility of forming rich and meaningful relationships,

- The potential to have complex emotions,

- An unparalleled ability to imagine a future and remember the past,

- A capacity for moral reasoning and

- A capacity to have faith and believe in a god (Savulescu 2003).

In addition, it is through the mental capacity of human beings that a person can understand concepts such as unconditional acceptance and religion and is able to confer human dignity upon himself or herself as well as upon others. In other words, if a human-nonhuman combination was created which undermined the distinction between the mental capacities of a human individual and a nonhuman animal then the very concept of full inherent dignity, such as human dignity, arising from these capacities could be challenged. However, the importance of intellectual abilities should not be over-emphasized. Indeed, many worldviews, including the Roman Catholic and many other Christian churches believe that persons with severe mental disorders are still endowed with the same moral status as that of a healthy human person even though they may have very limited mental capacity.

The transfer of brain cells between humans and nonhuman animals

Because of the profound relationship between the human brain and the sense of identity of a person, and because it is considered undesirable to alter nonhuman animals to the extent that they have a brain that would enable humanlike self-awareness, a number of recommendations have been proposed by the scientific community.

These include proposals from the previously mentioned 2005 inter-disciplinary working group in the USA which looked at research involving the engraftment of human neural stem cells into the brains of nonhuman primates. In this regard, it suggested the six following factors that research oversight committees should use as a starting framework (Greene et al. 2005):

- Amount/Proportion of engrafted human neuronal stem cells (ratio of foreign to endogenous cells at transplant),

- The stage of neural development of the prenatal or postnatal animal host,

- Species of the host animal,

- Brain size of the host animal,

- Specific site of integration into the animal brain of the human neural stem cells,

- Brain pathology (healthy or diseased) of the host animal prior to transplantation.

However, in the context of a proposed experiment in which human brain stem cells would be transplanted into foetal mice, another group led by Henry Greely of the Law School of Stanford University in the USA, indicated in 2002 that the following concerns should be taken into account (Greely et al. 2007):

1 The sources of the human brain stem cells,

2 The potential for pain and suffering to the mice,

3 The propriety of this use of human tissues (particularly brain tissues),

4 The risks of possibly conferring some degree of humanity on another species, and

5 The risks to public support of science.

Reproductive cells and human identity

Reproductive cells are strongly linked to human identity because they are instrumental to the manner in which a person is brought into existence. Moreover, the manner in which this is undertaken is vital to the manner in which the personal identity of a person is developed and accepted.

One of the defining attributes of being a human person has always been that the individual was procreated by other human persons giving him or her an origin and a history. In other words, a crucial aspect of the self-understanding and identity of a human being is given through knowing who his or her procreators really are and that they were human. This happens because there is 'something' from the procreators that is found in the created being. What this 'something' might be could often be hard to define but there is a sense that a person is entitled to human dignity because he or she is related to, and descendant from, all his or her historical predecessors who were also endowed with human dignity. In addition, there is undoubtedly a special relationship that comes into existence between the procreators and the procreated person because he or she arose from the procreators. Why this relationship is formed still requires further investigation.

This reality is also reflected in that significant interpersonal bonds of mutual belonging are often formed between the procreators, on the one hand, and the procreated person, on the other (MacKellar 1996). This is, for example, understood in the concepts that a person is accepted as a royal prince or princess because he or she was procreated by a king and/or queen or the ubiquitous phenomenon of having a family name as well as an individual given name. It is because a bond of mutual belonging is seen to exist between a son or daughter and his or her parents that the concept of inherited rights and identity is embraced by many societies.

In this context and if human-nonhuman combinations were ever born and became conscious, it would be very difficult for such a being to construct his or her identity and social identification. The individual may feel completely lost belonging neither to the human nor nonhuman worlds. Moreover, this is not only a problem for the individual, but has consequences for the whole of society. It may indeed be difficult for members of the general public to relate to the cross-species being with an indeterminate nature since they cannot place him or her in a clear position in society. As a consequence it may then be inevitable that his or her moral status would be undermined (Deutscher Ethikrat 2011, p. 62).

The transfer of reproductive cells between human and nonhuman animals

In the context of the creation of human-nonhuman combinations, a particular emphasis may be placed on those entities that are formed in the generative act itself such as in the creation of human-nonhuman hybrids through the fertilization of nonhuman eggs and human sperm. These entities would be created, from the outset, from human and nonhuman biological materials. This is in contrast to those entities created from already existing human and nonhuman animal life such as the creation of human-nonhuman chimeras through the amalgamation of early nonhuman and human embryos or through xenotransplantation.

With respect to the creation of human-nonhuman entities, the varying percentages of nonhuman or human genes in these new biological entities may be less relevant to their ethical status than the fact that they have been created by elements of two different species. The claim, for instance, that some hybrid embryos created through the use of nonhuman eggs and a human nucleus can be considered as 'human' because they will be 99.9 per cent human and 0.1 per cent nonhuman from a genetic perspective (Sample 2006) provides only a very incomplete description of the origins, biological characteristics and moral status that should be ascribed to these embryos. In this respect, the UK Academy of Medical Sciences in its 2007 report entitled 'Inter-Species Embryos' indicated that it was not the species membership per se of an entity that accounted for the dignity of a creature, an expression used to emphasize the special moral status and importance of human beings (AMS 2007: 29).

This suggests that it is not only the genetic material that matters but that other factors are also important. This includes the fact that a nonhuman egg was used in the creation of the embryo[4] since without this egg to reprogram the expression of genes in the transferred human nucleus, no living organism would ever be created. As indicated by the French embryologist Jean-Paul Renard, it is this important egg 'which will reorganize the nucleus back to an embryonic stage' (Renard 2002: 145).[5]

But whatever the procedure is considered, if an entity is accepted as having been (pro)created by human and nonhuman beings, then it is its whole identity and its entitlement to human rights and full inherent dignity that may be undermined and could be challenged.

---

[4]It can be also noted that human somatic cells are 100 per cent human but are not usually considered as having any special moral value.

[5]In fact, the mass of materials in the emptied animal egg is far greater than that of the injected human nucleus representing nearly 95 per cent of the total mass of the final entity (Deutscher Ethikrat 2011 p. 97).

Importance of the human gene

At this point, one could also point out that the genetic sequence that can be directly compared between the human and chimpanzee genomes is nearly 98 per cent identical. Even when DNA insertions and deletions are taken into account, humans and chimps still share 96 per cent of their sequence. At the protein level, 29 per cent of genes code for the same amino-acid sequences in chimps and humans (NIH News, 2005; Goodman, Grossman and Wildman 2005).Thus, most of the outwardly observable differences between the human and the chimpanzee result from different regulations of gene expression (Cobbe and Wilson 2011: 180).

A human being, moreover, generally only has two fewer chromosomes (46) than the chimpanzee's 48, with human chromosome 2 probably being a combination of two smaller chromosomes found in great apes such as chimpanzees, gorillas and orangutans. Does this then mean that great apes can also be considered as human (and vice versa)? Many of course would find difficulty in this conclusion. As the political scientist Andrea Bonnicksen notes 'human uniqueness cannot be established by genetic reference alone' (Bonnicksen 2009: 125).

In this regard, no agreement has even yet been reached concerning what is specifically human about the human genome. When the UNESCO was preparing its Universal Declaration on the Human Genome and Human Rights (1997), a number of definitions of the human genome were in fact proposed, including (Tonti-Filippini et al. 2006: 699):

- The genetic composition, as such, of human beings[6]

- The entire genetic material of humanity

- All the genes of every human being

- The tangible aspects of Human DNA and RNA molecules

- The immaterial aspects of human genetic information

- The genetic codes which are behind the vital functions of a human being

- The genetic material that can be detached from a human being

- The values and meanings attached to human identity.

---

[6]Interestingly, the Additional Protocol to the European Convention for the Protection of Human Rights and Dignity of the Human Being with regard to the Application of Biology and Medicine, indicates that 'the term human being "genetically identical" to another human being means a human being sharing with another the same nuclear gene set' (Council of Europe 1998a, Article 1).

Accordingly, it can be very difficult to define what is specifically human, as such, in a human genome or in human genes making it difficult to understand what is actually meant by a human-nonhuman interspecies being.

Thus, trying to define a specific feature that differentiates the human from the nonhuman is not as easy as may initially have been presumed. As the scientists Cobbe and Wilson indicate, 'If human identity depends merely on properties shared to some degree with other species, then determining whether or not some interspecies organisms should effectively be treated as members of the human species may be even more challenging' (Cobbe and Wilson 2011: 183). No one single factor in isolation makes human beings unique. Even differences based in intellectual capacities for reason and conscience are difficult to distinguish in the absence of any clear lines between the mental abilities of humans in comparison to those of other primates. A lot more work needs to be considered in this respect (Cobbe 2011: 129–56).

The international CHIMBRID project

As an application and example of the different extrinsic and intrinsic ethical concerns presented in the previous sections, the international CHIMBRID project may be considered. This was a project funded by the European Community which brought together 25 high-ranking scientists from 16 countries, and which sought to examine different approaches to the creation of the human-nonhuman combinations around the world while presenting various legislative and ethical settings. This endeavour also sought to find a proposed common ethical framework for considering such entities.

In its 2008 report, it suggested that the ethical aspects of interspecies entities could be discussed under the following headings: (I) Moral status, (II) Symbolic meaning, (III) Research ethics and (IV) Animal ethics. (Taupitz and Weschka 2009: 441–7). In its treatment of Animal ethics the report outlined the six moral considerations for animal welfare discussed above (p. 154). In relation to the question of the moral status of interspecies combinations, it is the first two which are the most relevant.

(I) Moral status

According to the CHIMBRID project, the first of these study levels, i.e. the moral status of an interspecies entity, could be considered from a further four different perspectives (Taupitz 2008; Fiester and Düwell 2009) which include both extrinsic and intrinsic ethical concerns and that generally reflect different properties of embryos and foetuses.

1. Ability to feel pleasure and pain

This relates to arguments associated with the extrinsic ethical concern of sentience. In this regard, it is suggested that the moral status of an entity is dependent on its capacity to feel pleasure and pain. Such a view would be one of the main considerations from a utilitarian perspective that does not generally recognize any idea of an inherent moral worth, which confers a special kind of dignity to human persons. In this case, concepts of rationality or self-consciousness are only relevant insofar as they can influence the capacity to feel pain and pleasure. From this viewpoint, the CHIMBRID project indicated that not many new ethical issues regarding moral status may be generated for early interspecies entities, such as embryos, since it is improbable that they could feel any pain (Fiester and Düwell 2009: 63).

2. Ability for self-consciousness and rationality

Alternatively, the moral status of interspecies entities could be examined by considering their ability to develop self-consciousness and rationality. This, as already mentioned, reflects the intrinsic ethical concerns associated with questions of dignity and personhood in the context of rationality and autonomy. In this case, it is possible to further distinguish between commentators who:

1   Consider the potential as sufficient for granting full moral status,

2   Consider the potential as sufficient to grant some moral status that   ·
is significantly different in comparison to the full moral status of persons,

3   Consider the end of pre-embryonic stage (14 days) of development as necessary to grant some or full moral status, and

4   Have a gradualistic position and believe that the actual capacity of an entity, such as an embryo or foetus, to live outside the mother, is crucial for its moral status.

In the first two points, the definition of 'potential' in the CHIMBRID report needs to be clarified since there are basically three ways of considering this concept:

a   Full moral status of the entity with the potential to develop.

First, it is possible to consider an entity, such as an embryo, as having a full moral nature or status as soon as it is created with the potential to continue in its development for a certain amount of time (be it short or long). This would mean accepting the need to fully respect and protect the embryo from the moment of its creation thus prohibiting embryo experimentation.

b The entity has the potential to develop into a being with greater moral status.

Under this concept, 'potential' represents the situation whereby the entity, such as an embryo, has a partial moral status, at most, but could develop into a being with a greater or full moral status.

In other words, the embryo could have some kind of special moral status depending on the stage of development and because it could grow into a being with greater or full moral status.

In this case, a gradualist approach could also be considered in which the moral status of a human-nonhuman embryo may increase in relation to, for example, its stage of development, its sensitivity, its number of human cells, etc. Note that the CHIMBRID analysis of gradualism implies that the protection afforded to entities before reaching full moral status can always be 'outweighed' by other important interests, such as the possibility of finding new treatments for medical disorders. They do not consider the possibility that less than full status might nevertheless be sufficient to give protection from deliberate destruction (given that other entities, such as animal species, that do not have the full moral status of persons may nevertheless be given at least protection against deliberate destruction).

c The moral status of the entity is uncertain.

In this third case, the concept of 'potential' reflects uncertainty regarding the exact moral status of an entity such as an embryo. Thus, while it may 'potentially' already be entitled to full moral status, it is impossible to determine this with any certainty.

3. Membership of the biological species Homo sapiens

Another manner of examining the moral status of human-nonhuman interspecies entity is to consider whether membership of the biological species *Homo sapiens* is a sufficient reason for granting moral status. This is related to previous discussions on the intrinsic ethical concern of speciesism. The CHIMBRID project mentioned that it is not clear why membership of one specific biological species should be a reason for granting a moral status

Nevertheless, this difficulty perhaps reflects an understanding of species as an abstraction rather than an actual existing community with a common life and a shared identity. Solidarity with other human beings is one of foundations for recognizing ethical status. This is not 'speciesist' because it does not deny that other species might also merit moral status.

4. Intrinsic value

Finally, the CHIMBRID project indicated that it is also possible to argue that an intrinsic value might be granted to sensitive beings, such as animals, which may be similar to the concept of animal telos, though the 'intrinsic value' term is not used in a clear and consistent manner. This would mean that, prima facie, animals would have some kind of value that makes it necessary to defend all kinds of behaviour that may harm this value. The intrinsic value of animals may thus be weighed against high moral goods, such as the expectation of valuable therapeutic outcomes for humans.

In conclusion, though recognizing that discrete boundaries for the moral status of interspecies entities may not exist, the CHIMBRID project suggested four broadly different positions in determining the moral status of human or human-nonhuman entities such as embryos (Fiester and Düwell 2009: 70–1). These were that moral status could be recognized:

> In entities that actually have interests. It is, therefore, not important whether an embryo is nonhuman or human or a combination of both. In other words, if the embryo is not sentient or does not have any interests, it need not be protected.

> In human entities, such as human embryos, from creation onwards. It is, therefore, extremely important to consider whether the embryo is nonhuman or human or a combination of both.

> In entities, such as embryos, but in different ways depending on whether they may develop into entities with full or only partial moral value.

> In entities, such as embryos, but in different ways depending on whether they have reached their fourteenth day of development.

(II) Symbolic meaning

The CHIMBRID project also recognized that, in considering the creation of human-nonhuman interspecies entities, there may be a 'symbolic dimension' which could relate to aspects, such as appearance, behaviour or emotion. This may reflect the discussion concerning the intrinsic ethical concern of unnaturalness mentioned previously. The project suggested that a symbolic order, that enables the recognition of humans as humans and nonhumans as nonhumans, may be important for humanity to acknowledge.

The project conceded that even though this dimension could be difficult to characterize, it could still be a basis for indicating that interspecies entities should not be created for arbitrary purposes. Indeed, rational agents could actually continue to argue that some elements in the natural world may have a symbolic importance in their understanding of themselves as free and rational beings (Fiester and Düwell 2009: 69).

Additional discussion

The report presented by the European CHIMBRID consortium can be recognized as a useful document in seeking to determine preliminary ethical aspects which should be considered in the creation of human-nonhuman interspecies entities. However, the report did not exhaust all the ethical considerations relevant to these entities. For example, the issue is understood primarily in relation to the protection of the creature that is conceived rather than the ethics of conceiving such a creature in the first place. This is almost certainly a mistake. The issue is not primarily whether human-nonhuman embryos deserve more protection than those of other animal species (though this is arguably the case). The primary issue is whether it is inherently unethical deliberately to create such entities. Furthermore, no real discussion was presented in the final analysis concerning the manner in which these entities could be considered from the perspective of inherent dignity. This oversight is unfortunate since it reduces the breadth of relevant arguments to be considered. The following chapter examining this concept of inherent dignity will, therefore, be of assistance in developing a deeper reasoning.

11

Ethical analysis

Within the animal kingdom, where does humanity begin? Where does it end? The countless myths portraying human-nonhuman combinations, throughout history, testify to the universality of these questions. Werewolves and fauns in Europe, dog-headed races in Africa, human-tigers in Asia, represent so many allegorical variations of this theme. These myths also question the very separation between the human and nonhuman species and the definition of humankind. In this regard, the French philosopher Pascal (1623–62) acknowledged that human beings were an enigma. Because of this, he advised that they should opt for religion instead of searching, in vain, for any certainties. But his compatriot, Voltaire (1694–1778), disagreed. For him (and in his characteristically cynical manner) humans were not a mystery but, rather, seemed to have their place in nature amongst the animals 'to which they are similar in their organs and … which they probably resemble in their thoughts' (Voltaire, XXV). A similar position was also reflected by Jean Bruller (1902–91), who considered human beings as denatured animals (in rebellion against nature). But he also believed that nonhuman animals were enslaved beings that could only attain 'personhood' in the awakening of their consciousness (Vercors 2009).

Understanding the difference and similarities between the human and nonhuman species has, however, become important since, as set out in the second part of this book, many kinds of human-nonhuman combinations are already possible or are likely to be a reality in the near future. Because of this, society is faced with a twofold challenge. In the first place, countries need to establish adequate legal categories to legislate for these developments. This is just as much of a challenge to permissive as it is to restrictive jurisdictions. Even in a permissive jurisdiction (such as the United Kingdom) it will be important, for example, for scientists to know whether the research counts as *human* fertility research or as nonhuman *animal* research, for these are covered by different laws and different systems of regulation. Secondly, societies need to come to a common ethical and political view about what action is necessary in this area. This is complicated by the fact that these are new and perplexing questions and how they are understood depends in some measure on the diverse worldviews that shape people's ethical understanding. Establishing a consensus in this area seems an impossible task.

From the preceding chapter the concept of full inherent dignity emerged as an important concept through which to analyse intrinsic concerns with human-nonhuman combinations. This chapter will, therefore, engage in a more detailed discussion of the application of this concept. The conclusion, which follows after this, will then explore some of the consequences of this discussion, and of other ethical considerations set out in the book as a whole, for the development of public policy on human-nonhuman combinations.

Full inherent dignity

As already mentioned, restricting the discussion relating to the dignity of human-nonhuman entities to human dignity, as such, may not be appropriate since many will only be partly human and to different degrees. In other words, it may be preferable to just use the term of full inherent dignity which is associate to having a full moral status. In this way, human dignity would only be one 'kind' of full inherent dignity which is associated with the species *Homo sapiens*.

It is also important to note that full inherent dignity, such as human dignity, has come under increasing attack lately with commentators, such as bioethicist Ruth Macklin and political theorist Alasdair Cochrane, suggesting that it is a meaningless concept with the latter arguing for an 'undignified bioethics' (Macklin 2003; Cochrane 2010). Cochrane suggests three reasons for this uselessness. First, human dignity is indeterminate in the sense that it very difficult to define or clarify (Macklin 2003). Secondly, it is often used in a reactionary manner as a moral trump card to limit advances in medicine (Pinker 2008). Finally, human dignity is considered to be redundant since it simply reflects a notion which is already emphasized in ethical principles and provisions, such as the respect for autonomy (Macklin 2003).

An example of such a perspective, which does not give a lot of importance to the concept of human dignity is the one found in the United Kingdom which has often been portrayed as having a 'pragmatic' approach to new scientific developments. This may be a consequence of its strong links with past and present utilitarian, including consequentialist, ideologies, an influence which may reflect the UK's important industrial heritage. In other words, new scientific possibilities have generally been welcomed in the UK if they proved useful, relatively safe and workable, this situation being made all the more possible because the concept of human dignity is not clearly recognized nor understood amongst the British general public. It is not considered as an objective notion and is even sometimes compared to the religious belief in the sanctity of human life which carries little weight in an increasingly secular society. Thus, the possibility of endangering the concept of human dignity is generally not considered as an important reason for restricting a useful procedure in the UK in contrast to a number of other European countries. Instead, analysis is confined to extrinsic concerns such as possible health risk weighed against the possible therapeutic or economic advantages of this avenue of research (Beyleveld, Finnegan and Pattinson 2009: 663).

Nevertheless, it should be noted that a society which no longer believes in the full inherent dignity of persons cannot offer any real argument against the taking of life of those who do not have certain characteristics, such as autonomy. It becomes a society that has lost its trust in the inherent value and meaning of life and cannot comprehend why it should be endured.

This is in complete opposition to a responsible benevolent and compassionate society which continues to affirm and defend the lives of all its members. It is also contrary to the notion that every human life is equal and full of value, meaning and richness even though persons may be aged, dependent on others or may have lost their autonomy.

The statement that inherent dignity, such as human dignity, is a useless concept would, finally, be in contradiction to the 1948 United Nations' Universal Declaration of Human Rights, which affirms in its preamble 'the inherent dignity and ... the equal and inalienable rights of all members of the human family' as 'the foundation of freedom, justice and peace in the world'.

This UN text emphasizes the universal, absolute, inalienable and inherent nature of the present concept of human dignity. It cannot be created, modified or destroyed by an individual, a majority or a state. To reject such a notion would not only seriously challenge the whole concept of inherent dignity but would be an extremely serious precedent in a world that has fought so hard to endow all persons with the same dignity. It is only because a concept, such as human dignity, is recognized as being extremely important that the rule of law can be maintained. Indeed, it is only because society believes in the full inherent dignity of persons, that it respects their autonomy.

Furthermore, the concept of inherent dignity cannot be reduced to any physical science since it is impossible to logically demonstrate that any beings, even rational and autonomous adult human beings, have any dignity, value or worth. It is simply something that is recognized or declared as in international declarations of human rights. In this respect, any attempt to measure inherent dignity constitutes a fundamental misunderstanding of the notion and risks undermining the concept by considering it as a tangible commodity. Societal progress requires the recognition that inherent dignity is a societal maxim that must be safeguarded and respected as inviolable for all persons. It is because of this concept that societies continue to protect those human beings who do not have any rational or autonomous capacities.

Nevertheless, it is still possible to try to address this perceived lack of clarity by considering the concept of full inherent dignity from two different perspectives. First, it may be seen from the perspective of the individual. This view requires others to respect and acknowledge a person's dignity. Secondly, it is possible to understand full inherent dignity as being grounded in a global network whereby every individual is both receiving and/or conferring this inherent dignity in a continuous fashion. This is a strong communal view of dignity. Indeed, the whole notion of dignity being inherent necessarily presupposes this global network since it is assumed that everyone in the network is entitled to this dignity which cannot be lost (De Melo-Martín 2008). From this perspective, it should be noted that though distinct, the two concepts are interdependent in that an individual would not enjoy the protection of inherent dignity without being in the receiving and/or conferring dignity network.

With this understanding of human dignity in mind, it is now possible to understand why the possible creation of human-nonhuman interspecies entities may undermine the very concept of full inherent dignity. First of all, new beings would begin to exist to whom/which it would be very difficult to ascertain, with any amount of certainty, whether or not universal, absolute, inalienable and inherent dignity should be conferred. In addition, if a being were denied the inherent dignity to which he or she was entitled, then the dignity of every individual in the whole global network, including every human being, would be put into question. This is because the whole network of persons (whether or not they are 100 per cent human) would no longer be consistent, coherent or dependable. The network would begin to be undermined together with the whole structure of society and its rule of law.

1. Questions relating to the inherent dignity of human-nonhuman entities

Until this present point in time, the specific inherent dignity particular to human dignity has always been a fact of humanity which has never really been questioned. There has always been a great moral divide between human beings and other animals even including nonhuman primates such as the great apes. The creation of human-nonhuman entities has, however, brought this fundamental distinction into question and, as such, marks a significant milestone in human history.

Though it is recognized that full inherent dignity cannot be diminished or taken away, it may be difficult, sometimes, to determine whether a being, such as a human-nonhuman combination, has this dignity. One view would only ascribe an interspecies entity full inherent dignity if it had the capacity for self-conscious rational thought. Indeed some suggest that inherent dignity cannot be reduced to the intrinsic characteristics of *Homo sapiens* but is related, instead, to a number of capacities such as reasoning or participating in social relations, not all of which need to be present in an entity (Karpowicz, Cohen and Van der Kooy 2005).

But determining whether an interspecies entity has these kinds of capacities which may then mean giving full moral status and protection may sometimes prove to be impossible since there are so many variables to be taken into account. Indeed many human beings may also not be autonomous or rational when a decision needs to be made and it may never be possible to determine, in any useful manner, whether they would ever eventually develop autonomous rationality (which for some religions may also mean after death).

Furthermore, every rational person is influenced by their culture and worldview which may be expressed in many different manners in the way in which they understand or apply ethical principles. Nevertheless, from this perspective, it is interesting to note that the overwhelming majority of commentators, including a large majority of the scientific community have never expressed a desire to create a baby with human and nonhuman parts (HFEA 2007a, p.12). The reason for this remains unclear but it seems that the birth of a creature with visible human and nonhuman features contradicts a very widely recognize sense of common humanity.

However, this reaction is less marked if the combination is not born but is destroyed early on in development, at the embryonic or foetal stage (though different people might begin to object at different stages of development and may also object to implanting a partly nonhuman creature in a woman, or implanting a partly human creature in a nonhuman animal). These differences seem superficially similar to differences of opinion about the status of the human embryo in relation to abortion (Jones 2010: 110). However, this is a very different issue. The issue is not primarily concerned with the protection of human-nonhuman combinations but with their creation. From this perspective, combinations at the origin of existence or at the early embryonic stage are *more* problematic than combinations that occur later. The earlier in development the combination occurs, the more tissue may eventually be involved and the deeper the degree of integration. This became evident in the discussion of different kinds of human-nonhuman combinations in Part II. It is most obvious when considering neural and germ cell lines, but it is also reflected in the degree of integration of the organism as a whole (as shown most clearly by true hybrids). Thus, while the emotional reaction to a human-nonhuman combination clearly varies with development, and some considerations (such as the psychological impact on the offspring) only apply later in development, the key intrinsic issue seems to be engaged from the moment in time that a combination is created.

In addition to the moment in time when an inter-species entity is generated where, paradoxically, earlier combinations are more troubling than later combinations, another important ethical consideration is the amount of time during which this entity is left to develop. For example, a number of different stages of development are already associated with a certain amount of protection to embryonic and foetal entities in certain countries, such as:[1]

Creation of the embryo,

The apparition of the primitive streak, corresponding to the end of individuation and the beginning of neurological development (14 days after creation for human embryos). This stage also relates to the upper limit of implantation into the mother's uterus of the embryo,

Half the gestation or incubation period for the relevant species has elapsed as defined, for example, in the UK Animals (Scientific Procedures) Act 1986,

The legal limit for abortions in many continental European countries (12 weeks after creation),

The last third of foetal development for the relevant species as defined, for example, in the 2010 EU Directive (2010/63/EU) on the protection of animals used for scientific purposes,

Birth of the infant at which stage most legislations confer full protection.

Moreover, the ethical evaluation of the combination also depends on the nonhuman species used (for example nonhuman primate), the nature of the biological material used (for example, embryonic or foetal material), the proportion of biological material used (whether significant in proportion to the recipient animal) and the integration sites of this material (for example, in the brain or some other part of the body). In this respect chimeras represent a more difficult challenge for legislators than other kinds of combination, because chimeras can be of very different proportions and raise very different issues. The previous chapter presented a rational justification for the special

---

[1] Relatedly, some commentators may include the possibility for an entity to develop from one moral status to the next. In other words, a 'gradual' position can exist in which an entity is entitled to an increasing degree of moral status in relation to its development. Degrees of moral status are thus acquired incrementally or when a specific stage of development is reached.

concerns that are frequently raised in relation to neurological and reproductive tissue. This is much less the case with other combinations, especially
where more distant species are used and where the research is justifiable
on other grounds. But trying to recognize the above features with respect to
a particular human-nonhuman entity may only represent some aspects in the
process of trying to determine whether it should be ascribed full inherent
dignity. Other aspects are also important such as the kind of relationships that
this entity may have with other beings that are already recognized to have this
full inherent dignity such as human beings. Thus, a comprehensive evaluation
relating to the different components of inherent dignity may be invaluable for
each different kind of human-nonhuman entity, though even this may still not
give any final answers. Nevertheless, if the propriety of some research in this
area is to be defended, a clear distinction between different cases must be
maintained. Effective legislation requires that clear lines are drawn.

From the previous discussion, the inevitable conclusion can only be that
the moral classification and appreciation of some human-nonhuman combinations will be extremely difficult. The main reason being that mixed creatures
do not fit easily into the existing categories of either human beings or
nonhuman animals with their respective rights. A certain amount of confusion
and disorder is then present. In this regard, it is possible to consider an interspecies entity in the following manners.

Firstly, it is possible to consider human-nonhuman combinations as having
no moral value or dignity. This may be because (1) the nature of the entity
created and/or (2) its stage of development does not enable the combination
to be self-aware or to support the concept of autonomy. Because of this, their
creation and destruction should not result in many new ethical problems.
Many of those supporting this position would also argue for a complete legislative deregulation in this area.

For those, however, who believe that human-nonhuman embryonic combinations cannot be assimilated to piles of cells, an alternative position would
be to give them a 'special' but not full moral status. In addition, if uncertainty
exists as to whether the entities should have a special moral status, some
may then decide that they should be given the benefit of the doubt.

This is also the position that considers the potential of the entity, such
as the embryo, to develop into something with moral value, given the right
nurturing, and that this has implications for the current actual status or
dignity of the entity. But this 'special' moral status is not often defined in any
clear way.[2] Generally, however, the mixing at a very intimate level of human

---

[2] It may be possible that some persons holding this position would eventually decide that the
biomedical benefits promised from the creation of human-nonhuman combinations compensate
the risks of destroying entities with a partial moral status.

and nonhuman biological material may be of concern to those who believe that this 'special' moral status can exist. This is because it would begin to undermine the whole distinction between human and nonhuman animals for which a different understanding of dignity exists.

Finally, it is also possible to consider human-nonhuman embryonic combinations as – or given the benefit of the doubt of – being persons endowed with the same moral value as other fully human persons. At present, it is virtually impossible for those holding this position to evaluate, a priori, whether a certain human-nonhuman embryonic or foetal combination should be considered as having full inherent dignity. This uncertainty would probably continue even if, for example, an embryonic or foetal combination was left to develop to term but, once born, remained unaware of its existence. However, if the entity was created then it is suggested that it should be given the benefit of the doubt and the same protection afforded to it as any other person in society. It would follow that destroying human-nonhuman combinations for research purposes would not be compensated by the promised biomedical benefits, but would be considered as a form of murder and be ethically indefensible.

2. Questions relating to the strength of a network of full inherent dignity

The philosopher, David DeGrazia argues that dignity or moral status would not be threatened by the creation of individuals, who may not be considered as belonging entirely to the human species but who still have full moral status. Thus, he adds that 'the transformation of … a Great Ape into a more humanlike person would not threaten human dignity' (DeGrazia 2007). He argues that either an interspecies entity would have dignity similar or identical to human dignity that other creatures lack or if they do not, they would not have this dignity. Either way, the distinction between creatures that possess full dignity and those that do not remains clearly defined as it is now.

In a similar vein, the UK Academy of Medical Science indicated in its 2007 report entitled 'Inter-Species Embryos' that if the concept of 'human dignity' had content, it is because there are factors of form, function or behaviour that confer such dignity or command respect. This means that society needs to decide whether interspecies creatures possess these factors which make them entitled to inherent dignity similar or identical to human dignity that other creatures lack or whether they do not (AMS 2007: 29).

In this regard, the Academy's response may be useful, if there are clear factors which can be used to determine with certainty between those beings that have and those that do not have this dignity. But as soon as there is doubt and hesitation – as soon as inherent dignity is not recognized and respected in individuals who may be entitled to such worth and value, or not given the benefit of the doubt, then real problems arise. This is because the concept of full and equal dignity is only protected if there are clear boundaries between those beings that do and those that do not have this dignity (Italian National Bioethics Committee 2009: 17). The importance of a clear demarcation or boundary between the human and nonhuman understanding of dignity was emphasized by Leon Kass, a former chair of the US President's Council on Bioethics, who indicated that:

> All of the boundaries that have defined us as human beings, boundaries between a human being and an animal on one side and between a human being and a super human being or a god on the other. The boundaries of life, the boundaries of death. These are the questions of the 21st century, and nothing could be more important. (Kass quoted in Smith 2005)

As already indicated, this boundary has until now been assumed to be a clear line that exists between the 'human' and all the other animal species. But the creation of human-nonhuman combinations may completely undermine this assumption. The term 'species' would then become far more challenging to define without any clear boundaries. As a result, the very concept of a 'human species' having fundamental consequences on the self-understanding, self-construction and self-perception of human beings would be undermined. The relationship between human beings and other life forms would have to be completely re-appraised in its most basic sense. As the ethicists Badura-Lotter and Düwell indicate 'it is hard to assume that we will be able to cope with the cataclysm that such a process would involve' (Badura-Lotter and Düwell 2009: 204).

An example of the problems that may arise when clear boundaries relating to dignity are not respected can be seen when the UK Human Fertilisation and Embryology Act 1990 was being prepared. Legislators at the time were anxious not to recognize full inherent dignity to the early human embryo, but neither were they prepared to completely deny this embryo any dignity. As a result, they decided to only give the human embryo a certain degree of this full dignity which reflected what they suggested was its 'special status' as characterized in the UK Warnock Report (Warnock 1984). But eventually, with the passing of the years, this concept of a special moral status of the embryo was all but abandoned. Without any clear definition, the concept was unintelligible and impracticable.

In December 2002, Baroness Warnock, admitted that:

> I regret that in the original report that led up to the 1990 legislation we used words such as "respect for the embryo" … I think that what we meant by the rather foolish expression "respect" was that the early embryo should never be used frivolously for research purposes'.

She added:

> You cannot respectfully pour something down the sink – which is the fate of the embryo after it has been used for research, or if is not going to be used for research or for anything else'. (*Lords Hansard*, Volume No. 641 Part No. 14, Column 1327)

More generally, it was noteworthy, that hardly any discussion relating to the special status of the human embryo remained during the preparation of the 2008 revision of the UK Human Fertilisation and Embryology Act 1990. The human embryo had, by then, gradually been reduced to a pile of cells in societal discussions with the next stage in the battle being the creation of human-nonhuman embryos.

This demonstrates that, even though full inherent dignity is impossible to completely define, an entity with a moral status which straddles two positions with respect to the boundary of full inherent dignity is ultimately untenable. Moreover, if there is any doubt or uncertainty in a significant number of persons or as soon as full inherent dignity is not conferred on individuals to whom it may be entitled, the whole concept of full dignity begins to be undermined since it becomes inconsistent and incoherent. This is because no clear agreement about the individuals on whom it should be conferred is possible.

As a consequence the global protective network of full inherent dignity becomes unstable. It would also mean that the foundational basis of the whole societal order would be brought into question since a clear demarcation of full inherent human dignity is essential for a civilized society to survive. This is why it is always appropriate to give entities the benefit of the doubt when their moral status is unclear or uncertain so that the strength of the concept and the network can be re-established. Indeed, the strength of the concept depends on it.

In doing so, the new uncertain frontier of full inherent dignity is thus widened to create a larger and stronger, border which includes all the beings of ambiguous moral status. The new boundaries will then become clear again to society. In other words, it is vital that any perception of the human-nonhuman interspecies beings should not bring about any perception of disorder, chaos or instability in society.

In the past and present, questions relating to assignment or conferment of full inherent dignity to individuals such as black people throughout the slave trade, indigenous Amerindians (raised during the Valladolid debate (1550–51)), Jews, non-white races, people with disabilities and a number of other minority groups in Nazi Germany, have all served to undermine the notion of full inherent dignity. Indeed it was this last and greatest violation of human dignity that inspired the UN Universal Declaration of Human Rights. In other words, as soon as this dignity is ascribed inconsistently and incoherently the very notion is undermined. This is because this dignity is no longer considered to be meaningful since it can eventually permit the killing and exploitation of certain categories of individuals. A key factor in these past controversies was the failure by society to accept the human dignity of these persons. Something that should have happened even though a large majority of the relevant society, at the time, was convinced of their lack of full inherent dignity. But because of the significant minority disagreeing strongly with this view, the majority should have given them the benefit of the doubt, a benefit that, if they were honest and consistent, they were also giving to themselves.

In other words, if some human-nonhuman interspecies entities are not given the full inherent dignity to which they are entitled, nor given the benefit of the doubt, this could eventually serve to challenge the very idea of conferring any kind of full inherent dignity to any individual – even those who may be completely human. The exploitation of interspecies entities would then not only offend the dignity due to the entities themselves but would also undermine the dignity of the whole community of individuals entitled to such a dignity (De Melo-Martín 2008). In this manner, the global protective network which this dignity gives would become unstable, inconsistent and unclear while the whole concept of society and its rule of law would begin to be injured. The social consequences resulting from the creation of beings of uncertain dignity should, therefore, not be underestimated. The very warning of the monstrous chimera symbolic, in ancient mythology, of some frightening disorder threatening both society and humanity with chaos may have been very appropriate.

In short, this means that full inherent dignity is not endangered just because human-nonhuman interspecies beings may not be part of the *Homo sapiens* species *per se* (Karpowicz, Cohen and Van der Kooy 2009: 546). Instead, it is the whole concept of conferring full inherent dignity and fundamental rights to all individuals within the global community, without having to pass some test of acceptance, that would be in danger of being undermined (Jones 2009c). The problem is that by creating human-nonhuman entities whom/which have an uncertain status as to their full inherent dignity, this would undermine the confidence of attributing full inherent moral status to many other beings (even if they are 100 per cent human) since the whole practice of conferring full inherent dignity would become unclear. As the German National Ethics Council indicated in its 2011 report, the birth of human-nonhuman entities with an unclear species membership will question whether they have an equal right to membership of human society. This is turn may threaten the very identity and uniqueness of the human kind as such (Deutscher Ethikrat 2011: 61).

These considerations do not, of course, apply equally to all human-nonhuman combinations. They do, however, imply that such entities should never be created unless they have a clear and uncontested moral status. If there is any uncertainty for a significant proportion of the general public, then their creation should not be permitted. Moreover, if these entities are unfortunately created with dubious moral status, then they should be given the benefit of the doubt as to their full inherent dignity.

3. Questions relating to the deliberate creation of entities with uncertain dignity

The act of deliberately creating an entity of unclear moral status by using human gametes, which are generally used to create beings with full inherent dignity, is also ethically problematic. This is because the use of such gametes in the creation of interspecies entities is a procedure that is similar enough to human procreation to reflect on the moral character of human procreation. Use of human gametes to create an interspecies entity would demonstrate a certain amount of indifference for the special importance human gametes have in the procreative process of humanity. Of course, it is recognized that human gametes are being discarded all the time. But this does not mean that when they are being used in a (pro)creative procedure they do not take on a special ethical and even philosophical meaning.

Indeed, if this special place of gametes were denied, they would then simply be reduced to that of mere biological matter. This would then reflect the position of a 2007 editorial of journal Nature, which indicated that '... biology's view of life as a molecular process lacking moral thresholds at the level of the cell is a powerful one' (Nature 2007).

The question here, however, is not whether gametes are entities deserving some level of protection. This is clearly not the case. Whereas embryos are generally recognized as having an inherent moral status (though it is disputed whether this is *full* inherent status). Gametes do not have any inherent moral status. Their significance is in relation to their use. Human dignity is important here not in relation to protecting cells but in relation to the process of procreation. This is not only reflected in various religious worldviews but seems to express a common human understanding, as implied by the common legal prohibition against intercourse with a nonhuman animal. It is for the same reason that 'true hybrids' are sometimes put into a separate category to other human-nonhuman combinations.

12

Conclusion

uman beings are in a privileged position since they can intellectually and morally understand themselves in relation to their nature and to recognize the conditions which they required for the preservation and development of their existence. They are also the only animals capable of being aware of their dependence on other organisms and, as such, have a special responsibility towards the manner in which to draw the natural, historical and cultural foundations for life. However, the manner in which humankind understands and defines itself is dependent on the existence of a clear line determining humanity from other animals.[1] This then serves not only to oppose the special moral status of humanity to the rest of the animal word but to also repudiate certain nonhuman animal characteristics and behaviours in its search for a specific identity (Kollek 2011). The line then forms the basis of moral and legal differences but also plays a crucial and meaningful role in religious and cultural foundations (Deutscher Ethikrat 2011: 9).

Because of this, and as already indicated in the legal section of this book, a number of policies and regulations have been suggested in the area of embryonic, foetal and post-natal human-nonhuman combinations to continue to uphold this important demarcation line. These originate from university ethics committees, to government funding agencies and finally to legislation. In preparing these guidelines it is important that these bodies do not restrict potentially useful and significant biomedical research without good reason. However, it is also essential to restrict and in some cases to prohibit certain procedures where these give rise to well-recognized ethical concerns. Research, like all worthwhile human activities, is worthy of respect only if it is

---

[1] The apparently clear ethical and legal demarcation between human and nonhuman animal rights is already beginning to be permeable. This is especially the case with primates which are increasingly given a certain measure of human rights and created human embryos for research for whom human rights are being challenged (Kollek 2011, p. 128).

conducted within ethical limits. It should never undermine the special inherent dignity of animals or human beings. In this regard, the US neurologist William Hurlbut warns, in an article entitled 'The Boundaries of Humanity', that 'we must be concerned about our slow but steady drift towards treating all of living nature (including human nature) as mere matter and information to be reshuffled and reassigned for projects of the human will' (Hurlbut 2011: 168).

Some of the consequences of the preceding ethical reflections for the development of public policy on human-nonhuman combinations are set out below. These will draw not only on one principle but on a number of principles and aspects discussed through the book. Proposals will be all the stronger when they reflect a convergence of different ethical concerns, and it is such proposals that have the best hope of promoting a strong ethical consensus.

This last section is structured around the recommendations of Scottish Council on Human Bioethics (SCHB) from its report of 2006, which was the first report on the topic from an ethics council in Europe (Scottish Council on Human Bioethics 2006). These still provide a valuable orientation in this area and their influence may be detected at a number of points through this book.

Necessity for public discussion

The discussion of cultural perspectives in Chapter 8, and the detailed account given in various places in the book of the UK debate prior to the Human Fertilisation and Embryology Act 2008 shows the real need for well-informed public discussion on these issues. In a democratic society and on a question that involves fundamental questions of humanity, it is necessary that a wide range of people contribute to the development of public policy. The ethical issues in relation to human-nonhuman combination are genuinely difficult and to make progress as a society a number of insights are necessary which are drawn from different worldviews, both from a diverse range of secular worldviews and from religious traditions (which remain an important source of ethical reflection for many people). In this respect the right starting point would seem to be the encouragement of public reflection and open consultation by the appropriate bodies, which was supported by the SCHB in 2006:

> National Ethics Committees of the Council of Europe member states should initiate, as soon as possible, an extensive consultation and reflection relating to the complex ethical questions arising from the creation of human-nonhuman combinations.

> National governments should ensure that the fundamental questions raised by the developments of biology and medicine are the subject of

appropriate public discussion in the light, in particular, of relevant medical, social, economic, ethical and legal implications, and that their possible application is made the subject of appropriate consultation (Council of Europe 1997, Article 28).

The Parliamentary Assembly and the Steering Committee on Bioethics of the Council of Europe should address the ethical issues arising from the creation of human-nonhuman combinations, as soon as possible, in a Recommendation and/or a legally binding Convention.

Nevertheless, as highlighted in the section on secular worldviews, it is useful to ask whether governmental and other bodies may be unduly influenced by economic or other interests in a way that could distort the process of consultation or public engagement. In the light of the criticisms levelled at the HFEA consultation on hybrid embryos in the UK (Baylis 2009), it should be added that for such consultation to be genuine and democratic in spirit it must occur before policy has been agreed. It is all too easy for a process of 'consultation' to degenerate into an exercise in persuasion or of permission after the fact.[2] Thus, it is important not only that such consultations occur but that they are robustly challenged to demonstrate their openness to diverse perspectives.

Clarity in legislation

The legal discussion in the first part of the book and the considerations in the scientific and ethical parts each confirmed the importance of clarifying the manner in which certain human-nonhuman combinations will be regulated. In particular it is essential to know whether proposals fall within dedicated legislation, within existing animal-welfare legislation, within other legislation, or outside legislation altogether. Even within the United Kingdom that invested considerable parliamentary time in 2008 into clarifying the different categories of 'human admixed embryo', there remains legal uncertainty about human-nonhuman combinations where nonhuman DNA predominates. Hence 'the UK legislative structure is such that some awkward cases may fall at the boundary of jurisdiction' (AMS 2011: 113). There is a need within all jurisdictions to review their current legislation in order to help identify the gaps and boundary issues that require attention. This need has not abated since it was identified by SCHB in 2006:

---

[2]This is not a new phenomenon, as was wittily observed by Ambrose Bierce in his *Devil's Dictionary* who defined consult as follows: 'CONSULT, v.i. To seek another's approval of a course already decided on'.

In so far as it is possible, a decision should be taken to determine whether a created human-nonhuman entity should come under human or nonhuman animal legislation.

Another theme that emerged, both in the context of the discussion of legislation, and from the ethical analysis, was the need for some 'bright lines' to provide ethical orientation in this area. Without some clear prohibitions in place as a set of secure reference points it is difficult for systems of regulation or ethics committees to maintain ethical consistency. It is also the case that, all other things being equal, there is a great advantage if support can be given to existing international reference points. It was noted above that the European Convention on Human Rights and Biomedicine (Council of Europe 1997) is one of the most important legally binding instruments in existence in Europe on bioethical issues. This convention does not prohibit the use of human embryos or cells derived from human embryos but it does prohibit the creation of human embryos purely for the purpose of research. This prohibition gives content to the idea of 'respect for the embryo' in contrast to the 'special status' of the embryo which ostensibly provides an ethical basis for UK legislation but which is now widely regarded as an empty concept. In relation to convergence, the prohibition of creating embryos for research also undermines the trade in human eggs for research, prevents research that would facilitate human reproductive cloning and restricts at least some forms of genetic modification of human embryos. It is for these and other reasons that the United States federal government and the European Union will not give funding to the creation of embryos for research.

It is unclear whether the prohibition of creating human embryos for research (Council of Europe 1997, Article 18.2) also covers human(-nonhuman) admixed embryos. However, discussion of the intrinsic concerns raised by human-nonhuman combinations shows that these are greater the earlier in development the combination occurs. If it is ethical and prudent, for a broad range of reasons, to support the European Convention on Human Rights and Biomedicine and prohibit the creation of human embryos for research, there is still more reason to support a prohibition on the creation of human admixed embryos. This recommendation of SCHB is one of the bright lines that needs to be drawn:

The creation of human [admixed] embryos for research purposes, using human and nonhuman biological material, should be prohibited. [This corresponds to Council of Europe 1997, Article 18.]

Human-nonhuman transgenesis

The creation of human-nonhuman transgenic entities is an area that has been examined for some time, in research. In this regard, the SCHB recommended that:

> The creation of transgenic nonhuman animals in which some foreign human genes are deliberately inserted into the genome of nonhuman animals should only proceed with extreme caution.

> Somatic gene therapy interventions in which some foreign nonhuman animal genes are deliberately inserted into the genome of human beings should only be undertaken for preventive, diagnostic or therapeutic purposes and only if its aim is not to introduce any modifications in the genome of descendants. [This corresponds to Council of Europe 1997, Article 13.]

It is true that transgenesis does not necessarily raise the question of perplexity in the way that a true hybrid does but this is one area where those putting together this volume were not able to find a consensus. If a human protein is expressed by a single celled organism, this does not generally mean that the organism has become human. On the other hand the use of higher animals raises welfare concerns and it is not clear where to draw the line in relation to the influence of multiple genes or genes that may have multiple uses. It is noteworthy in this context to remember that transgenic combinations can arguably be understood as a kind of hybrid, and part of that continuum. The SCHB statement would also not go far enough for some people as it implicitly allows interventions that knowingly introduce nonhuman genes into the germline (if this is not one of the 'purposes' of the intervention).

Human-nonhuman gestation

No regulations or national legislation anywhere in the world have supported the possibility of human-nonhuman gestation. This would be unethical for a great number of reasons from both intrinsic and extrinsic perspectives. There is, therefore, no reason for anyone to resist the proposals of SCHB that:

> The placing of a live human embryo into a nonhuman animal should be prohibited.

The placing of live human sperm into a nonhuman animal should be prohibited.
The placing of a live nonhuman embryo into a woman should be prohibited.
The placing of live nonhuman sperm into a woman should be prohibited.

[This corresponds to Council of Europe 1986b, Article 14 (A).]

Human-nonhuman hybrids

The mixing of human and nonhuman gametes represents the most extreme and most perplexing form of human-nonhuman combination. It is also the form that is closest to human procreation. In the case of use of human ova it also raises the danger of the abuse of women. In addition to the ethical concerns already mentioned in the previous chapters, it falls foul of the general prohibition, defended above, of creating embryos for research purposes and there is no clear scientific benefit from doing this. Thus, even if it were not ruled out by all the preceding reasons, it would fail the test of proportionality. There are convergent reasons to prohibit the creation of hybrids and, notwithstanding the general prohibition of creating admixed embryos, it is worth making this a separate prohibition, as this would be likely to attract more widespread support. Linked to this should be a prohibition on creating embryos containing both human and nonhuman chromosomes as this would be similar in effect to the mixing of gametes, even if it were achieved by a different method. Thus:

The mixing of human and nonhuman gametes should be prohibited.

The creation of an embryo possessing cells containing both human and nonhuman chromosomes should be prohibited.

Human-nonhuman cytoplasmic hybrids

Similar concerns in relation to true hybrids also apply to the creation of hybrids by somatic cell nuclear transfer. In this case the danger of facilitating human reproductive cloning is a further reason for prohibition. As argued above, this technology also threatens to have a negative impact on women, even if nonhuman ova are used in the first instance. It is part of a wider research project to obtain stem cells from human embryos cloned using human eggs. The debates surrounding the UK Human Fertilisation and Embryology Act

2008 demonstrated how support for cybrids established a precedent for the acceptance of true hybrids. This seems to show a slippery slope between cybrids and other forms of hybrids, which is a further reason to draw a line with the prohibition of the creation of admixed embryos. Hence for convergent reasons the creation of human-nonhuman cytoplasmic embryos also merits to be proscribed. Thus:

> The insertion of a human cell nucleus or chromosomes into a nonhuman egg stripped of its chromosomes enabling an embryo to exist should be prohibited.

> The insertion of a nonhuman cell nucleus or chromosomes into a human egg stripped of its chromosomes enabling an embryo to exist should be prohibited.

Human-nonhuman chimeras

The creation of chimeras represents the most subtle and difficult issue within the area of human-nonhuman combinations. Some forms of chimera, such as those involved in xenotransplantation do not necessarily raise issues of perplexity or human dignity. They do not necessarily involve processes in any way similar to procreation and, in the case of human cells transplanted into nonhuman animals, do not necessarily give rise to human characteristics at the level of the organism as a whole. On the other hand, in the preceding chapter it was argued that neurological or reproductive tissue belong to a different category precisely because these parts of the animal represent the whole in some sense. Furthermore, interventions earlier in development are more problematic in part because they produce combinations which could affect neurological or reproductive function and characteristics (but also because they represent a more fundamental level of integration). The ethical conclusions of this discussion give good reason, therefore, to support the recommendations of the SCHB in this area:

> Xenotransplantation should only take place if the procedure respects all national and international legal instruments such as the Council of Europe Recommendation (2003) 10 of the Committee of Ministers on Xenotransplantation.

> The incorporation of human stem cells into post-natal nonhuman animals should proceed with extreme caution. Moreover, such a procedure should only take place if it can be demonstrated that the cells cannot contribute

to the germline or give rise to specifically human brain functions in the nonhuman animals.

The incorporation of human stem cells into post-blastocyst stages of nonhuman embryos should only take place if it can be demonstrated that they cannot contribute to the germline or brain cells of the nonhuman animal.

The incorporation of nonhuman stem cells into post-blastocyst stages of human embryos should only take place if it can be demonstrated that they cannot contribute to the germline or brain cells of the human being.

The incorporation of human pluripotent or totipotent stem cells into a nonhuman blastocyst or its preliminary embryonic stages should be prohibited.

The incorporation of nonhuman pluripotent or totipotent stem cells into a human blastocyst or its preliminary embryonic stages should be prohibited.

Fostering further reflection

This last chapter has applied the considerations of the book to certain definite proposals in order to bring into focus the discussion of the book as a whole. While there are different aspects to this issue: historical, legal, scientific, cultural, worldview and ethical, these aspects all need to inform public policy. Nevertheless, while some concrete proposals have been defended, the aim of this book has not been primarily to establish one or other of these proposals. Rather, the aim has been to show the range of aspects that need to be taken into account in thinking about this issue, and the need for a greater range of voices from a greater range of perspectives to engage with the questions.

Thus many questions remain relating to ascribing moral status to the different human-nonhuman entities being suggested. Of course, it may be possible to argue that there would be no ethical concerns (other than individual biomedical and psychological risks) with the creation of human-nonhuman combinations if all biological species were freely accorded full and equal rights. However, society does not do this, and it is difficult to evaluate a priori the kind of dignity and hence the extent to which specific protective rights should be accorded to new human-nonhuman combinations. Thus, society at present can consider the following three options:

First, it can create certain kinds of human-nonhuman combinations and allow them to develop to term. However, if they then prove to be associated with biomedical developmental disorders, psychological problems or societal inequalities and prejudice, society might eventually decide that they should never have been created in the first place.

Secondly, it can create certain kinds of human-nonhuman combinations and kill them before they develop to any advanced stage because society cannot, as yet, deal with the ethical issues which they pose. In this case, there is also the risk of killing a living entity, out of prejudice, which may be entitled to full inherent dignity.

Finally, it can decide not to create certain kinds of human-nonhuman combinations because it cannot, as yet, deal with the ethical and societal consequences and instability which they initiate. Indeed, in some cases, it seems that a certain verdict or answer on the moral nature of some created human-nonhuman entities is virtually impossible (Tonti-Filippini et al. 2006: 696). Thus, if the moral status of a human-nonhuman combination cannot be determined without creating such an entity, that in itself should then be a sufficient argument against its creation.

Glossary

blastocyst a hollow ball of 50 to 100 cells which is obtained after about four to five days of human embryonic development just before implantation in the uterus.

blastomere a single cell in an embryo just after fertilization.

cell line cells of common descent and type cultured in the laboratory.

cell nuclear replacement (also called somatic cell nuclear transfer) the procedure of replacing the cell nucleus of an egg with the nucleus from another cell.

cell type one of over 200 different types of cells in the human body, for example blood cells, liver cells, neural cells. Each of these cell types has a different subset of genes switched on ('expressed') and therefore specific characteristics which allow these cells to serve a specific function in the body.

chimera an organism composed of cells derived from at least two genetically different cell types. The cells could be from the same or separate species.

clone a cell or organism derived from and genetically identical to another cell or organism.

cytoplasm a gel-like substance, which together with the nucleus which it surrounds, forms the interior of the cell.

dedifferentiation the process of inducing a specialized cell to revert towards pluripotency.

differentiation the process by which less specialized cells develop into more specialized cell types.

DNA deoxyribonucleic acid – the cell's and the body's genetic material.

ectoderm the outermost of the three primitive germ layers of the embryo; it gives rise to skin, nerves, and brain. It appears about 16 days after fertilization for the human embryo.

embryo (human) a human being during the first 56 days of his or her development following fertilization or creation, excluding any time during which the development has been suspended.

endoderm the innermost of the three primitive germ layers of the embryo; it later gives rise to the lungs, liver and digestive organs. It appears about 16 days after fertilization for the human embryo.

enucleated from which the nucleus has been removed (usually of an egg).

gamete the male sperm or female egg.

gastrulation the procedure by which an animal embryo at an early development stage produces the three primary germ layers ectoderm, mesoderm, and endoderm. This takes place about 16 days after fertilization for the human embryo.

genome the complete genetic material or makeup of an individual.

***in vitro* fertilization** the fertilization of an egg by a sperm outside the body.

mesoderm the middle layer of the embryonic disk, which consists of a group of cells derived from the inner cell mass of the blastocyst; it is formed at gastrulation and is the precursor to bone, muscle and connective tissue. It appears about 16 days after fertilization for the human embryo.

mitochondria sometimes described as 'cellular power plants', because they convert organic materials into energy. They do not form part of the nucleus (containing the chromosomes) of a cell.

morula a solid mass of 16–32 cells resulting from the cell division of a zygote (a fertilized egg).

multipotent having the capacity to develop into multiple (but not all) cell types.

oocyte immature female egg.

ovum (ova) mature female egg (eggs).

pluripotent having the capacity to develop into every cell type in the human body, but not the extra-embryonic tissues such as the placenta and the umbilical cord.

primates group of mammals which include monkeys, apes and humans.

primitive streak appears around the fourteenth day after creation of the embryo. This is the latest stage at which identical twins can occur.

The streak also includes precursor cells to the spinal cord.

redifferentiation the process of inducing a dedifferentiated cell to differentiate into a (different) specialized cell type.

replication the production of two identical DNA molecules from an original DNA molecule.

spermatid immature spermatozoon.

spermatozoon (spermatozoa) male germ cell (cells).

stem cell a cell that has the ability to divide for an indefinite period *in vivo* or in culture and to give rise to specialized cells.

totipotent having the capacity to develop into every cell type required for human development, including extra-embryonic tissues.

trophectoderm the extra-embryonic part of the ectoderm of mammalian embryos at the blastocyst stage before the mesoderm becomes associated with the ectoderm.

transcription the synthesis of RNA under the direction of DNA.

xenotransplantation the transplantation of cells, tissue or organs from a donor of one species into a recipient of another species.

zygote the single cell formed when the male sperm fertilizes the female egg.

Bibliography

Legislation and Official Reports

AMRC (2008), *Briefing from the UK Association of Medical Research Charities*, March 2008

AMS (2007), *Inter-species Embryos* (London: Academy of Medical Sciences)

—(2011), *Animals Containing Human Material* (London: Academy of Medical Sciences)

Animal Procedures Committee (2001), *Report on Biotechnology* (London: Home Office)

—(2003), *Review of Cost-Benefit Assessment in the Use of Animals in Research* (London: Home Office)

Animals (Scientific Procedures) Act 1986

BACS (2008), 'Human-Animal Combinations for Biomedical Research: A Consultation Paper', *Bioethics Advisory Committee of Singapore*, 8 January 2008

Banner, M. C. (1995), *Report of the Committee to consider the ethical implications of emerging technologies in the breeding of farm animals* (London: HMSO)

Canadian Assisted Human Reproduction Act 2004

CDF (1987), *Instruction on respect for human life in its origin and on the dignity of procreation, replies to certain questions of the day* (Vatican: Congregation for the Doctrine of the Faith)

—(2008), *Instruction Dignitas Personae on Certain Bioethical Questions* (Vatican: Congregation for the Doctrine of the Faith)

CIHR (2010), *Updated Guidelines for Human Pluripotent Stem Cell Research. June 30, 2010* (Canadian Institutes of Health Research) available at http://www.cihr-irsc.gc.ca/e/42071.html

Commission of the European Communities (2000), 'Communication from the Commission on the precautionary principle' *Commission of the European Communities*, 2 February 2000

Council of Europe (1986a), *European Convention for the Protection of Vertebrate Animals used for Experimental and other Scientific Purposes*, CETS No.: 123 (Strasbourg: Council of Europe)

—(1986b), *Parliamentary Assembly of the Council of Europe, Recommendation 1046 (1986) on the use of human embryos and foetuses for diagnostic, therapeutic, scientific, industrial and commercial purposes* (Strasbourg: Parliamentary Assembly of the Council of Europe)

—(1987), *European Convention for the Protection of Pet Animals*, CETS No.: 125 (Strasbourg: Council of Europe)

—(1997), *Convention for the Protection of Human Rights and Dignity of the Human Being with regard to the Application of Biology and Medicine: Convention on Human Rights and Biomedicine*, CETS No.: 164 (Oviedo: Council of Europe)

—(1998a), *Additional Protocol to the Convention for the Protection of Human Rights and Dignity of the Human Being with regard to the Application of Biology and Medicine, on the Prohibition of Cloning Human Beings*, CETS No.: 168 (Paris: Council of Europe)

—(1998b), *Protocol of Amendment to the European Convention for the Protection of Vertebrate Animals used for Experimental and other Scientific Purposes*, CETS No.: 170 (Strasbourg: Council of Europe)

—(2003), *Recommendation Rec(2003)10 of the Committee of Ministers to member states on xenotransplantation*, Adopted by the Committee of Ministers on 19 June 2003 at the 844th meeting of the Ministers' Deputies (Strasbourg: Council of Europe)

—(2005), *Parliamentary Assembly of the Council of Europe: Embryonic, Foetal and Post-natal Animal-Human Mixtures, Doc. 10716, 11 October 2005, Motion for a resolution presented by Mr Wodarg and others* (Strasbourg: Parliamentary Assembly of the Council of Europe)

Department of Health (2000a), *Stem cell research: Medical progress with responsibility A Report from the Chief Medical Officer's Expert Group reviewing the potential of developments in stem cell research to benefit human health* (London: Department of Health)

—(2000b), *Government response to the recommendations made in the Chief Medical Officer's Expert Group Report: 'Stem Cell Research: Medical Progress with Responsibility'* (London: Department of Health)

Deutscher Ethikrat (2011), *Mensch-Tier-Mischwesen in der Forschung: Stellungnahme* (Berlin: Deutscher Ethikrat)

DGHC (2000), 'Press release – Commission adopts Communication on Precautionary Principle', *Directorate General for Health and Consumers*, 2 February 2000

The Embryo Protection Act (*Embryonenschutzgesetz*) 13 December 1990 (passages quoted translated by C. MacKellar)

EPO (2010), *European Patent Office Guidelines for Examination* (Munich: European Patent Office)

EU Directive 86/609/EEC *COUNCIL DIRECTIVE of 24 November 1986 on the approximation of laws, regulations and administrative provisions of the Member States regarding the protection of animals used for experimental and other scientific purposes* (Brussels: The Council of the European Communities)

—98/44/EC of the European Parliament and of the Council of 6 July 1998 on the legal protection of biotechnological inventions, *Official Journal of the European Union* L 213: 0013–21

—2010/63/EU of the European Parliament and of the Council of 22 September 2010 on the protection of animals used for scientific purposes, *Official Journal of the European Union* L 276: 33–79

French National Consultative Bioethics Committee (1999), *Opinion on Ethics and Xenotransplantation* N°61 (Paris: Comité Consultatif National d'Éthique pour les sciences de la vie et de la santé)

Guidance on the Operation of the Animals (Scientific Procedures), Act 1986

Hansard, (the Official Report) available at http://www.parliament.uk/business/publications/hansard/

HCSTC (2005), *Human Reproductive Technologies and the Law*, House of Commons Science and Technology Committee Fifth Report of Session 2004–5, Vol. I (London: HMSO)

—(2006), *Scientific Advice, Risk and Evidence Based Policy Making*, House of Commons Science and Technology Committee Seventh Report of Session 2005–6, Vol. I (London: HMSO)

—(2007), *Government Proposals for the Regulation of Hybrid and Chimera Embryos* House of Commons Science and Technology Committee No. 28A of Session 2006–7 (London: HMSO)

HFEA (2006), *Scientific Horizon Scanning at the HFEA, Annual Report 2006* (London: Human Fertilisation and Embryology Authority)

—(2007a), *Hybrids and Chimeras: A Consultation on the Ethical and Social Implications of Creating Human/Animal Embryos in Research, April 2007* (London: Human Fertilisation and Embryology Authority)

—2007b), *A report on the findings of the consultation Hybrids and Chimeras, October 2007* (London: Human Fertilisation and Embryology Authority)

House of Lords (HL) (2002), Report from the Select Committee of the House of Lords, Stem Cell Research, 200

Draft Human-Animal Hybrid Prohibition Act of 2007 (S.2358)

Human-Animal Hybrid Prohibition Act of 2008 (H.R. 5910)

Human-Animal Hybrid Prohibition Act of 2009 (S. 1435)

Draft Human Chimera Prohibition Act of 2005 (S.1373)

ISSCR (2006), *Guidelines for the conduct of human embryonic stem cell research Version 1: December 21, 2006* (Deerfield, IL: International Society for Stem Cell Research)

Italian National Bioethics Committee (2009), *Chimeras and Hybrids with Specific Attention to Cytoplasmic Hybrids*, 26 June 2009 (Rome: Italian National Bioethics Committee)

Joint Committee (2007), *Report of the Joint Committee on the Human Tissue and Embryos (Draft) Bill* Volume I, HL Paper 169–I and HC Paper 630–I

Louisiana Senate Bill No. 115 (Act 108)

Lowell v. Lewis, 15 (a. 1018 No. 8568) (C.D. mass. 1817)

National Academy of Sciences (2005), *Guidelines for Human Embryonic Stem Cell Research* (Washington, DC: National Academies Press)

—(2010), *Final Report of The National Academies' Human Embryonic Stem Cell Research Advisory Committee and 2010 Amendments to The National Academies' Guidelines for Human Embryonic Stem Cell Research* (Washington, DC: National Academies Press)

NIH (2009), *National Institutes of Health Guidelines for Human Stem Cell Research* (Bethesda, MD: National Institutes of Health)

Nuffield Council on Bioethics (2003), *The Use of Genetically Modified Crops in Developing Countries* (London: Nuffield Council on Bioethics)

—(2005), *The Ethics of Research Involving Animals* (London: Nuffield Council on Bioethics)

—(2007), *The forensic use of bioinformation: ethical issues* (London: Nuffield Council on Bioethics)

People Science & Policy (2008), *Public Attitudes to Science 2008: A survey* (London: Research Councils UK / Department for Innovation, Universities and Skills)

Polkinghorne, J. ed. (1993), *Report of the Committee on the Ethics of Genetic Modification and Food Use*, Ministry of Agriculture, Fisheries and Food (London: HMSO Publications)

Pontifical Academy for Life (2001), *Prospects for Xenotransplantation – Scientific Aspects and Ethical Considerations* (Vatican: Pontifical Academy for Life)

Preamble (1948), *Preamble to the United National Declaration of Human Rights*

President's Commission (1983), *President's Commission for the Study of Ethical Problems in Medicine and Biomedical and Behavioral Research, Deciding to Forgo Life-Sustaining Treatment* (Washington, DC: US Government Printing Office)

The President's Council on Bioethics (2002), *Human Cloning and Human Dignity* (Washington, D.C. The President's Council on Bioethics)

—(2004), *Reproduction and Responsibility: The Regulation of New Biotechnologies* (Washington, D.C. The President's Council on Bioethics)

Prohibition of Human Cloning for Reproduction Act 2002

Prohibition of Human Cloning for Reproduction and the Regulation of Human Embryo Research Amendment Act 2006

R (Quintavalle) v Secretary of State for Health [2003] UKHL 13

Rio Declaration on Environment and Development (1992), The United Nations Conference on Environment and Development, Rio de Janeiro, 3–14 June 1992

Scottish Council on Human Bioethics (2006), 'Embryonic, Fetal and Post-natal Animal-Human Mixtures: An Ethical Discussion', *Human Reproduction and Genetic Ethics* 12 (2): 35–60

Sexual Offences Act 2003

South Korea: Bioethics and Biosafety Act (2004), (unofficial translation available at www.ruhr-uni-bochum.de/kbe/Bioethics&BiosafetyAct-SouthKorea-v1.0.pdf)

TCPS (2005), Canadian Institutes of Health Research, Natural Sciences and Engineering Research Council of Canada, and Social Sciences and Humanities Research Council of Canada, *Tri-Council Policy Statement: Ethical Conduct for Research Involving Humans*, August 1998 with 2000, 2002 and 2005 amendments

—(2010), Canadian Institutes of Health Research, Natural Sciences and Engineering Research Council of Canada, and Social Sciences and Humanities Research Council of Canada, *Tri-Council Policy Statement: Ethical Conduct for Research Involving Humans*, December 2010

The UABResearch Foundation, T. M. Townes and T. Ryan, 2000 Patent Application WO/2000/069268: 'Production of Human Cells, Tissues, and Organs in Animals', International Filing Date: 12 May 2000

UNESCO (2009), *Report of the International Bioethics Committee on Human Cloning and International Governance* SHS/EST/CIB-16/09/CONF.503/2 Rev

Warnock M. (chair) (1984), *Report of the Committee of Enquiry into Human Fertilisation and Embryology* (London: Her Majesty's Stationery Office)

Further bibliography

Abbott, A. and Cyranoski, D. (2001), 'China plans 'hybrid' embryonic stem cells', *Nature* 413: 339

Andorno, R. (2001), 'The Paradoxical Notion of Human Dignity', *Rivista Internazionale di filosofia del diritto* 78: 151–68

—(2004), 'The Precautionary Principle: A New Legal Standard for a Technological Age', *Journal of International Biotechnology Law* Vol 01: 1

Anon. (1925), 'Hybridization of Man and Ape To Be Attempted in Africa', *Daily Science News Bulletin*, No. 248 D. E. F., sheets 1–2, in Rossiianov 2002: 295

—(1926), 'Savant to Try Hybridization of Man and Ape: Plans Complete for Experiment in Africa', *Des Moines Sunday Register*, 3 January 1926, in Rossiianov 2002: 295

APF (2008), 'New S Korean law tightens rules on cloning', *SEOUL (AFP)*, 16 May 2008

Aquinas, Thomas, *Summa theologiae*, (1272), (trans.) English Dominican Fathers (1948) (New York, NY: Benziger)

Aristotle, *Generation of Animals*, (trans.) A. L. Peck (1953) (London: Loeb Classical Library).

—*History of Animals: Books VII–X*, (trans.) D. M. Balme (1991) (London: Loeb Classical Library).

The Associated Press (2001), 'Stem cells may help in brain repair', *USA Today: Health & Science* 26 July 2001

—(2005), 'Scientists create animals that are part-human', *Associated Press* 29 April 2005

Atkins, M. (2000), 'Three Ways to Love an Animal' New Blackfriars 81 (950): 108–123

Attari, G. and A. Chomyn (1995), *Methods in Enzymology Volume 260: mitochondrial biogenesis and genetics Part A* (New York: Academic Press)

Augustine, *City of God* (1984), (trans.) H. Bettenson (London: Penguin books)

Bader, M. (2009), 'Transgenic animals carrying human genes: Methods and Ethical Aspects', in Taupitz and Weschka 2009

Badura-Lotter, G. and M. Düwell (2009), 'Chimeras and Hybrids – How to Approach Multifaceted Research?' in Hug and Hermerén 2011: 203–4

Balaban, E. (1997), 'Changes in multiple brain regions underlie species differences in a complex, congenital behavior', *Proc Natl Acad Sci USA* 94 (5): 2001–6

Batty, D. (2007), 'Q&A: Hybrid embryos', *Guardian*, 17 May 2007

Bayertz, K. ed. (1996), *Sanctity of Life and Human Dignity* (Dordrecht: Kluwer)

Baylis, F. (2008), 'Animal Eggs for Stem Cell Research: A Path Not Worth Taking,' *American Journal of Bioethics* 8(12): 18–32

—(2009), 'The HFEA Public Consultation Process on Hybrids and Chimeras: Informed, Effective, and Meaningful?' *Kennedy Institute of Ethics Journal* 19 (1) March 2009: 41–62

Baylis, F. and C. Mcleod (2006), 'Feminists on the Inalienability of Human Embryos' *Hypatia* 20 (1): 1–14

Bazopoulou-Kyrkanidou, E. (2001), 'Chimeric creatures in Greek mythology and reflections in science', *Am. J. Med. Genet.* 100: 66–80

BBC News Online (1998), 'Company "cloned human cells"', 13 November 1998, http://news.bbc.co.uk/1/hi/sci/tech/213663.stm

—(1999), 'Details of hybrid clone revealed', 18 June 1999, http://news.bbc. co.uk/1/hi/sci/tech/371378.stm

—(2008), 'UK's first hybrid embryos created', 1 April 2008, http://news.bbc. co.uk/1/hi/health/7323298.stm

BBVA (2008), *Second BBVA Foundation International Study on Biotechnology: Attitudes to Stem Cell Research and Hybrid Embryos* (Bilbao: Fundación BBVA)

Beauchamp, T. and L. Walters (eds) (1994), *Contemporary Issues in Bioethics, 4th edn* (Belmont, MD: Wadsworth)

Beazley, C.R. ed. (1903), *The Texts and Versions of John de Plano Carpini and William de Rubruquis as Printed for the first time by Hakluyt in 1598 together with some shorter pieces* (London: The Hakluyt Society)

Beckmann, J. P. (1998), '*Patientenverfügungen: Autonomie und Selbstbestimmung vor dem Hintergrund eines im Wandel begriffenen Arzt-Patient-Verhältnisses*', *Zeitschrift für Medizinische Ethik* 44

Bedford, J. M. (1977), 'Sperm/egg interaction: the specificity of human spermatozoa', *Anat Rec.* 188:477–87

Behringer, R. (2007), 'Human-Animal Chimeras in Biomedical Research', *Cell Stem Cell*, 1 (3): 259–62

Bentham, J. (1823), An Introduction to the Principles of Morals and Legislation

Berg, T. (2006), 'Human Brain Cells in Animal Brains – Philosophical and Moral Considerations', *The National Catholic Bioethics Quarterly*, Spring 2006

Beyleveld, D. and R. Brownsword (2001), *Human Dignity in Bioethics and Biolaw* (New York: Oxford University Press)

Beyleveld, D., T. Finnegan and S. D. Pattinson (2009), 'The Regulation of Hybrids and Chimeras in the UK', in Taupitz and Weschka 2009

Bhan, A., P. Singer and A. S. Daar (2010), 'Human-animal chimeras for vaccine development: an endangered species or opportunity for the developing world?', *BMC International Health and Human Rights* 10: 8

Birnbacher, D. (1996), 'Ambiguities in the concept of *Menschenwürde*', in Bayertz 1996: 91–106

Boethius (1918), *Tractates and the consolation of philosophy* (trans.) H. F. Stewart, E. K. Rand and S. J. Tester, (Cambridge, MA: Harvard University Press)

Bonnicksen, A. L. (2009), *Chimeras, Hybrids and Interspecies Research: Politics and Policymaking*, (Washington, D.C: Georgetown University Press)

Bowring, F. (2003), *Science, Seeds and Cyborgs: Biotechnology and the Appropriation of Life*, (London: Verso)

Boyce, N. (2003), 'Mixing species – and crossing a line?', *usnews. com*, 27 October 2003, http://www.usnews.com/usnews/culture/ articles/031027/27chimeras.htm

Bradshaw, P (2010), 'Splice: Film Review', *Guardian*, 22 July 2010

Brom, F. W. A. and E. Schroten (1993), 'Ethical questions around animal biotechnology: the Dutch approach', *Livestock Production Science* 36: 99–107

Brown, G. (2008), 'Why I believe stem cell researchers deserve our backing', *Observer: Comment* 18 May 2008

Burke, D. (2006), 'Ethics in Public Policy', *Presentation given to the Faraday Institute in Cambridge*, 24 September 2006

Butler D. (1998), 'Last chance to stop and think on risks of xenotransplants', *Nature* 391:320–26

Byrne, J. A. et al. (2003), 'Nuclei of adult mammalian somatic cells are directly reprogrammed to oct-4 stem cells gene expression by amphibian oocytes', *Current Biology* 13: 1206–13

Campbell, A. (2009), *The Body In Bioethics*, (London: Routledge-Cavendish)

Casabona, C. M. R. and I. de M. Beiain (2009), 'Legal regulation in different countries' in Taupitz and Marion 2009: 367–78

Chung, Y. et al. (2009), 'Reprogramming of Human Somatic Cells Using Human and Animal Oocytes', *Cloning and Stem Cells*, 11(2)

Clarke: and A. Linzey (1988), *Research on Embryos: Politics, Theology and Law* (London: Lester Crook Academic Publishing)

Clayton, D. A. et al. (1971), 'Mitochondrial DNA of human-mouse cell hybrids', *Nature* 234: 560–62

Cobbe, N. (2007), 'Cross-Species Chimeras: Exploring a Possible Christian Perspective', *Zygon: Journal of Religion and Science* 42: 599–628

—(2011), 'Interspecies mixtures and the status of humanity' in Huarte and Suarez 2011

Cobbe, N. and V. Wilson (2011), 'Creation of Human–Animal Entities for Translational Stem Cell Research: Scientific Explanation of Issues That Are Often Confused', in Hug and Hermerén 2011

Cochrane, A. (2010), 'Undignified bioethics', *Bioethics* 24 (5): 234–41

Coghlan, A. (2003), 'First human clone embryo ready for implantation', *NewScientist.com* 15 September 2003, http://uk.news.yahoo.com/030916/12/e8k6h.html

Collingwood, R. G. (1960) (1945), *The Idea of Nature*, (Oxford: Oxford University Press)

Crichton, M. (2006), *Next*, (New York: Harper Collins)

Cyranoski, D. (2008), '5 things to know before jumping on the iPS bandwagon', *Nature* 452, 27 March 2008

Dawkins, R. (1976), *The Selfish Gene*, (Oxford: Oxford University Press)

Deacon, T. et al. (1997), 'Histological evidence of fetal pig neural cell survival after transplantation into a patient with Parkinson's disease', *Nature Medicine* 3: 350–3

De Francesco, L. et al. (1980), 'Uniparental propagation of mitochondrial DNA in mousehuman cell hybrids', *Proc Natl Acad Sci USA* 77: 4079–83

DeGrazia, D. (2007), 'Human-Animal Chimeras: Human Dignity, Moral Status, and Species Prejudice', *Metaphilosophy* 38: 309–29

De Melo-Martín, I. (2008), 'Chimeras and Human Dignity', *Kennedy Institute of Ethics Journal*, 18(4): 331–46

De Pomerai, D. (1997), 'Are there Limits to Animal Transgenesis?', *Human Reproduction and Genetic Ethics*, 3(1)

De Sousa, R. (1980), 'Arguments From Nature', *Zygon* 15 (2): 169–91

—(1984), 'The Natural Shiftiness of Natural Kinds', *Canadian Journal of Philosophy* 14: 561–81

Dickenson, D. (2006), 'The Lady Vanishes: What's Missing from the Stem Cell Debate,' *Journal of Bioethical Inquiry* 3: 43–54

Dominko, T. et al. (1999), 'Bovine oocyte cytoplasm support development of embryos produced by nuclear transfer of somatic cell nuclei from various mammalian species', *Biol. Reprod.* 60(6): 1496–1502

Eberl, J. T. (2005), 'Aquinas's account of human embryogenesis and recent interpretations', *Journal of Medicine and Philosophy* 30(4): 379–94

—(2006), *Thomistic principles and bioethics*, New York, NY: Routledge

—(2007), 'Creating Non-Human Persons: Might it be worth the risk?', *American Journal of Bioethics*, 7(5): 52–4

Eberl, J. T., and R. A. Ballard (2009), 'Metaphysical and Ethical Perspectives on Creating Animal-Human Chimeras', *Journal of Medicine and Philosophy* 34(5): 470–86

Emborg, M. E. et al. (2008), 'GDNF-secreting human neural progenitor cells increase tyrosine hydroxylase and VMAT2 expression in MPTP treated cynomolgus monkeys', *Cell Transplant* 17: 383–95

Ereshefsky, M. (2007), 'Species', in E. N. Zalta ed. *The Stanford Encyclopedia of Philosophy* http://plato.stanford.edu/archives/sum2007/entries/species

Fehilly, C. B. and S. M. Willadsen (1986), 'Embryo manipulation in farm animals', *Oxford Reviews of Reproductive Biology* 8: 379–413

Fiester, A. and M. Düwell (2009), 'Ethical Issues Raised by Chimeras and Hybrids – An Overview', in Taupitz and Weschka 2009: 61–77

Gardner, R. L. (1968), 'Mouse chimeras obtained by the injection of cells into the blastocyst', *Nature* 220: 596–7

George, K. (2008), 'Women as Collateral Damage: a Critique of Egg Harvesting for Research,' *Women's Studies International Forum* 31(4) *Special Issue: Women and Technologies of Reproduction* 285–92

Gewirth, A. (1978), *Reason and Morality* (Chicago: University of Chicago Press)

Giles, R. E. et al. (1980), 'Characterization of mitochondrial DNA in chloramphenicol-resistant interspecific hybrids and a cybrid', *Somatic Cell Genet* 6: 543–54

Godlovitch, S., R. Godlovitch and J. Harris (eds) (1971), *Animals Men and Morals* (London: Victor Gollancz)

Goldstein, R. S., M. Drukker, B. E. Reubinoff and M. Benvenisty (2002), 'Integration and differentiation of human embryonic stem cells transplanted to the chick embryo', *Developmental Dynamics* 225 (1): 80–6

Goodman, E. (2001), 'Scientists, ethicists cautiously give nod to Zanjani's research', *Reno Gazette-Journal*, 22 October 2001

Goodman, M., L. I. Grossman and D. E. Wildman (2005), 'Moving primate genomics beyond the chimpanzee genome', *Trends in Genetics*, 2005, 21 (9): 511–17

Greely, H. T. (2003), 'Defining chimeras ... and chimeric concerns', *American Journal of Bioethics* 3:17–20

—et al. (2007), 'Thinking About the Human Neuron Mouse', *American Journal of Bioethics* 7(5): 27–40

Greene, M. et al. (2005), 'Moral Issues of Human-Non-Human Primate Neural Grafting', *Science* 309 (5733): 385–6

Griffiths, D. J. et al. (2002), 'Novel endogenous retrovirus in rabbits previously reported as human retrovirus', *Journal of Virology* 76: 7094–102

Haddow, G., A. Bruce, J. Calvert, S. H. E. Harmon and W. Marsden (2010), 'Not "human" enough to be human but not "animal" enough to be animal – the case of the HFEA, cybrids and xenotransplantation in the UK', *New Genetics and Society* 29 (1): 3–17

Hagen, G. R. and S. A. Gittens (2008), 'Patenting Part-Human Chimeras, Transgenics and Stem Cells for Transplantation in the United States, Canada, and Europe', *Richmond Journal of Law & Technology* 14(4)

Harmon, S. H. E. and N.-K. Kim (2008), 'Medical Research Governance in Korea: The New Bioethics and Biosafety Amendment Bill (Draft 17-8353) or "Inertia Reiterated"', *SCRIPTed* 5(3): 575–82

Harris, J. (1999), 'The concept of the person and the value of life', *Kennedy Institute of Ethics Journal*. 9(4): 293–308

Hartman, T. and R. Williams (1993), 'The ethics of species manipulation', *Science and Christian Belief* 5: 117–37

Hauskeller, M. (2005), 'Telos: the Revival of an Aristotelian Concept in Present Day Ethics', *Inquiry* 48 (1): 62–75

Haworth, J. (2008), 'Should we beware of the apeman's coming?', *The Scotsman*, 29 April 2008

Heeger, R. and F. W. A. Brom (2001), 'Intrinsic value and direct duties: from animal ethics towards environmental ethics?', *Journal of Agriculture and Environmental Ethics* 14: 241–52

Henderson, M. (2004), 'Human-animal cell experiments fall outside law', *The Times*, 1 June 2004

—(2007), 'Medicine faces ban on rabbit-human embryos', *The Times*, 5 January 2007

—(2008a), 'Scientists win public support on embryo research', *The Times*, 10 April 2008

—(2008b), 'MPs back creation of human-animal embryos', *The Times*, 20 May 2008

—(2009), 'Human-animal hybrid embryos: the debate revisited a year on', TimesOnline, 29 June 2009

Highfield, R. (2001), 'How to make babies without a man', *Telegraph*, 10 July 2001

Hilli, A. al- (no date), *Al-bab al-hadi al-ashar* (Qum: Maktab Intisharat)

Hinsliff, G. (2008), 'Brown says embryo research is key to life', *Observer*, 18 May 2008

Hitler, A. (1939) (1927), *Mein Kampf* trans. J. Murphy (London: Hurst and Blackett)

Holland, A. (1990), 'The Biotech Community: A Philosophical Critique of Genetic Engineering', in Wheale and McNally 1990

Holland, S. (2001), 'Contested commodities at both ends of life: Buying and selling gametes, embryos, and body tissues' *Kennedy Institute of Ethics Journal* 11 (3): 263–84

Homer, *The Iliad,* (1924) (trans.) A. T. Murray (London: Loeb Classical Library)

Huang, S.-Z. et al. (2006), 'Multiorgan engraftment and differentiation of human cord blood CD34+Lin– cells in goats assessed by gene expression profiling', *PNAS* 103(20): 7801–6

Huarte, J. and A. Suarez (2011), *Is this Cell a Human Being?: Exploring the Status of Embryos, Stem Cells and Human-Animal Hybrids* (Berlin; New York: Springer)

Hug, K. (2009), 'Research in Human-animal Entities: Ethical and Regulatory Aspects in Europe', *Stem Cell Rev and Rep* 5: 181–94

Hug, K. and G. Hermerén (eds) (2011), *Translational Stem Cell Research, Stem Cell Biology and Regenerative Medicine* (Dordrecht, The Netherlands: Springer)

Hughes, J. (2004), *Citizen Cyborg: Why Democratic Societies Must Respond to the Redesigned Human of the Future* (Boulder, Colorado: Westview Press)

Hurlbut, W. (2011), 'The Boundaries of Humanity: The Ethics of Human-Animal Chimeras in Cloning and Stem Cell Research' in Huarte and Suarez 2011

Hyun I. et al. (2007), 'Ethical Standards for Human-to-Animal Chimera Experiments in Stem Cell Research', *Cell Stem Cell* 1: 159–63

Ikumi, S. et al. (2004), 'Interspecies somatic cell nuclear transfer for in vitro production of Antarctic minke whale (*Balaenoptera bonaerensis*) embryos', *Cloning Stem Cells* 6(3): 284–93

Illmensee, K., M. Levanduski and P. M. Zavos (2006), 'Evaluation of the embryonic preimplantation potential of human adult somatic cells via an embryo interspecies bioassay using bovine oocytes', *Fertility and Sterility*, 85 (supplement 1): 1248–60

Ipsos MORI (2010), *Exploring the Boundaries, Report on a public dialogue into Animals Containing Human Material*, September 2010, http://www.ipsos-mori. com/DownloadPublication/1377_sri-exploring-the-boundaries-ipsos-mori-ams-september-2010.pdf, Accessed 14 December 2010

James, D., S. A. Noggle, T. Swigut and A. H. Brivanlou (2006), 'Contribution of human embryonic stem cells to mouse blastocysts', *Developmental Biology*, 295(1): 90–102

Jasanoff, S. (2005), *Designs on nature: science and democracy in Europe and the United States* (Princeton: Princeton University Press)

Johnston, J. and C. Eliot (2003), 'Chimeras and "Human Dignity"', *American Journal of Bioethics* 3 (3): 6–7

Jones, D. A. (2004), *The Soul of the Embryo: An enquiry into the status of the human embryo in the Christian tradition* (London: Continuum)

—(2005), 'The appeal to the Christian tradition in the debate about embryonic stem cell research' *Islam and Christian-Muslim Relations* 16 (3)

—(2009a), 'What Does The British Public Think About Human-Animal Hybrid Embryos?' *Journal of Medical Ethics* 35: 168–70

—(2009b), 'Nothing but a sideshow' *Guardian: Comment is Free*, 16 January 2009

—(2009c), 'Incapacity and personhood: Respecting the non-autonomous self' in Watt 2009

—(2010), 'Is the Creation of Admixed Embryos 'an offence against Human Dignity'? *Human Reproduction and Genetic Ethics* 16(1): 87–114

Jones, N. L. (2003), *Could Animal-Human Chimeras Be On the Way?* (Deerfield, IL: Centre for Bioethics and Human Dignity)

Kaneda, H. et al. (1995), 'Elimination of paternal mitochondrial DNA in intraspecific crosses during early mouse embryogenesis', *Proceedings of the National Academy of Sciences USA* 92: 4542–6

Kant, I. (1964) (1785), *Groundwork of the Metaphysics of Morals* (trans.) H. J. Paton (New York: Harper Torchbooks)

Karpowicz, P., C. B. Cohen and D. J. Van der Kooy (2004), 'It is ethical to transplant human stem cells into nonhuman embryos', *Commentary, Nature Medicine*, 10 (4): 331–35

—(2005), 'Developing Human-Nonhuman Chimeras in Human Stem Cell Research: Ethical Issues and Boundaries', *Kennedy Institute of Ethics Journal* 15(2): 107–34

—(2009), 'Developing Human-Nonhuman Chimeras in Human Stem Cell Research – Ethical Issues and Boundaries', in Taupitz and Weschka 2009

Kashani, M. al-Fayad (1980) (1358 HS), *'Ilm al-yaqin fi usul al-din'* (Qum: Intisharat Bidar)

Kass, L. (1985), *Towards a More Natural Science: Biology and Human Affairs* (New York: The Free Press)

—(1998), 'The Wisdom of Repugnance' in J. Wilson ed., *The Ethics of Human Cloning* (Washington DC: American Enterprise Institute)

Kemp: (1998), 'Four Ethical Principles in Biolaw', *2nd International Conference on Bioethics and Biolaw*, Copenhagen, 3–6 June 1998

Keown, J. and D. A. Jones (2008), 'Surveying the Foundations of Medical Law: A Reassessment of Glanville Williams's The Sanctity of Life and the Criminal Law', *Medical Law Review* 16(1): 85–126

Kobayashi, T. et al. (2010), 'Generation of Rat Pancreas in Mouse by Interspecific Blastocyst Injection of Pluripotent Stem Cells', *Cell* 142(5): 787–99

Kollek R. (2011), 'Sondervotum' in Deutsche Ethikrat, 126–35

Lacham-Kaplan, O., R. Daniels and A. Trounson (2001), 'Fertilization of mouse oocytes using somatic cells as male germ cells', *Reproductive Biomedicine Online* 3: 205–11

Lawler, R., J. Boyle and W. E. May (1985), *Catholic Sexual Ethics: A Summary, Explanation, & Defense*, (Huntington, IN: Our Sunday Visitor)

Leader, D. P. (2003), 'Reproductive cloning: an attack on human dignity', *Nature* 424: 14

Leake, J. (2003), 'Cloning expert claims to have created 'human-cow' embryo', *Times Online*, 14 September 2003

Leake, J. and S.-K. Templeton (2008), 'Mice produce human sperm to raise hope for infertile men', *Sunday Times*, 6 July 2008

Lee, B. et al. (2003), 'Blastocyst development after intergeneric nuclear transfer of mountain bongo antelope somatic cells into bovine oocytes', *Cloning Stem Cells* 5(1): 25–33

Lee, R. G. and D. Morgan (2001). *Human Fertilisation & Embryology: Regulating the Reproductive Revolution* (London: Blackstone Press)

Leroi, A. M. (2005), *Mutants: On the Form, Varieties and Errors of the Human Body* (London: Harper Perennial)

Levin, I. 1991 (1976), *The Boys from Brazil* (New York: Bantam)

Li, Y. et al. (2006), 'Cloned endangered species takin (*Budorcas taxicolor*) by inter-species nuclear transfer and comparison of the blastocyst development with yak (*Bos grunniens*) and bovine', *Molecular Reproduction and Development* 73(2): 189–95

Locke, J. (1975) (1690), *An essay concerning human understanding* (Oxford: Clarendon Press)

Logan, J. S. and A. Sharma (1999), 'Potential use of genetically modified pigs as organ donors for transplantation into humans', *Clinical and Experimental Pharmacology and Physiology* 26(12): 1020–5

Loike, J. D. and M. Tendler (2008), 'Reconstituting a Human Brain in Animals: A Jewish Perspective on Human Sanctity', *Kennedy Institute of Ethics Journal* 18 (4): 347–67

Lu, F. et al. (2005), 'Development of embryos reconstructed by interspecies nuclear transfer of adult fibroblasts between buffalo (*Bubalus bubalis*) and cattle (Bos indicus)', *Theriogenology* 64(6): 1309–19

MacKellar, C. (1996), 'The Biological Child', *Ethics and Medicine* 12(3): 65–9

Macklin, R. (2003), 'Dignity is a Useless Concept', *British Medical Journal* 327: 1419–20

Malpas, J. and N. Lickiss (eds) (2007) *Perspectives on Human Dignity: A Conversation*, (Dordrecht, The Netherlands: Springer)

Marcus Aurelius, *Meditations* (2003), (trans.) G. Hays (New York: Random House Modern Library)

Matthews, E. and M. Menlowe (eds) (1992), *Philosophy and health care* (Brookfield, VT: Avebury)

McLaren, A. (1976), *Mammalian Chimeras* (Cambridge: Cambridge University Press)

—(2007), 'Free-Range Eggs?', *Science* 20 April 2007, 316 (5823): 339

Meilander, G. (2007), 'Human Dignity and Public Bioethics', *The New Atlantis* 17 (Summer 2007): 33–52

Meinecke-Tillmann, S. and B. Meinecke (1983), 'Possibilities and limits of micromanipulation of the stages of embryonic division in domestic animals demonstrated by an artificial monozygotic twin model in sheep', *Zentbl. VetMed.* 30: 146–53

Mench, J. A. (1999), 'Ethics, animal welfare and transgenesis', in Murray, J. D. et al. *Transgenic animals in Agriculture*, CAInternational

Midgely, M. (2000), 'Biotechnology and Monstrosity; Why We Should Pay Attention to the "Yuk Factor"', *Hastings Center Report*, 30 (5) (September–October 2000): 7–15

—(1983), *Animals and Why They Matter: a journey around the species barrier* (Harmondsworth: Penguin Books)

Mill, J. S. (1987) (1863), *Utilitarianism* (Prometheus Books, Buffalo, NY)

Minger, S. (2006), 'Junk medicine: therapeutic cloning', *The Times*, 11 November 2006

Misselbrook, D. (2004), 'Speciesism', *Christian Medical Fellowship Files*, No. 26

Mitchell, C. B. et al. (2007), *Biotechnology and the Human Good* (Washington, D.C.: Georgetown University Press)

Modell, S. M. (2007), 'Approaching Religious Guidelines for Chimera Policymaking', *Zygon: Journal of Religion and Science* 42 (3): 629–42

Moens, H. M. B. (1908), *Truth: Experimental Researches about the Descent of Man* (London: A. Owen)

Muotri et al. (2005), 'Development of functional human embryonic stem cell-derived neurons in mouse brain', *Proceedings of the National Academy of Science of the United States of America* 102 (51): 18644–8

Murakami, M. et al. (2005), 'Development of interspecies cloned embryos in yak and dog', *Cloning Stem Cells* 7(2): 77–81

Nagy, A. et al. (1993), 'Derivation of completely cell culture-derived mice from ear passage embryonic stem cells?' *Proceedings of the National Academy of Sciences of the United States of America* 90 (18): 8424–8

Nasr, S. H. (1993), *An Introduction to Islamic Cosmological Doctrines* (Albany, NY: State University of New York Press)

Nature, (2007), 'Editorial: meanings of "life": synthetic biology provides a welcome antidote to chronic vitalism', *Nature* 447: 1031–2

Nau, J.-Y. (2007), '*Cellules souches: le Royaume-Uni autorise des chimères d'humain et d'animal*', *Le Monde* 22 May 2007 (passages quoted translated by C. MacKellar)

New Scientist, (2003), ' "Humanised" organs can be grown in animals', *New Scientist*, 17 December 2003

—(2004), 'Pig-human chimeras contain cell surprise', *New Scientist*, 13 January 2004

Nicholson, J. K. et al. (2004), 'The challenges of modelling mammalian biocomplexity', *Nature Biotechnology* 22: 1268–74

NIH News (2005), 'New Genome Comparison Finds Chimps, Humans Very Similar at the DNA Level', *NIH News* 31 August 2005

N.R.D. (2009), 'Patent Watch', *Nature Reviews Drug Discovery* 8 (1): 12–13

O'Doherty, A. et al. (2005), 'An aneuploid mouse strain carrying human chromosome 21 with Down syndrome phenotypes', *Science* 23 September 2005, 309(5743): 2033–7

O'Donovan, O. (1985), *Begotten or Made* (Oxford: Oxford University Press)

Palmiter, R. D. et al. (1982), 'Dramatic growth of mice that develop from eggs microinjected with methallothionein-growth hormone fusion genes', *Nature* 300: 611–15

Panjwani, S. and I. Panjwani (2010), *Islamic Metaphysics in Bioethics: Animal-Human Experimentation*, (London: Centre for Islamic Shi'a Studies Press)

Pascal, B., *Thoughts* (1910), ed. C. W. Eliot, (trans.) W. F. Trotter, Harvard Classics, vol. 18 (New York: P. F. Collier & Son)

Pearsall, J. and B. Trumble (eds) (1996), *The Oxford English Reference Dictionary, Second Edition*, (Oxford: Oxford University Press)

Peters, T., K. Lebacqz, and G. Bennett (2010), *Sacred Cells? Why Christians Should Support Stem Cell Research* (Lanham, MD: Rowman & Littlefield Publishers)

Pinker, S. (2008), 'The Stupidity of Dignity', *New Republic*, 28 May 2008

Polzin, V. J. et al. (1987), 'Production of Sheep-Goat Chimeras by Inner Cell Mass Transplantation', *Journal of Animal Science* 65: 325–30

Powell, A. (2007), 'Stem cells, through a religious lens', *Harvard News Office*, 22 March 2007

Primrose, S. B., R. Twyman and B. Old (2001), *Principles of Gene Manipulation, 6th edition* (Malden, MA: Blackwell Science)

Rabderson, J. and R. Dawkins (2009), 'How would you feel about a half-human half-chimp hybrid?', *Guardian*, 2 January 2009

Rahamni, S. (1997), (1376HS) *Tajalli wa – zuhur dar irfan nazari* (Qum: Markaz Intisharate Daftar Tablighat Islami)

Rahman, F. (1975), *The Philosophy of Mulla Sadra* (Albany: State University of New York Press)

Rauber, A. (1886), '*Personaltheil und Germinaltheil des Individuum*', *Zool Anz* 9: 166–71

Redmond Jr., D. E. (2002), 'Cellular replacement therapy for Parkinson's disease – Where we are today?', *Neuroscientists* 8 (5): 457–88

Redmond Jr., D. E. et al. (2010), 'Cellular repair in the parkinsonian nonhuman primate brain' *Rejuvenation Res* 13(2–3):188–94

Renard, J.-P. (2002), '*Le Clonage*' in Y. Michaud 2002. *Universite de Tous les Savoirs, Vol. 4: La Vie*, (Paris: Poches Odile Jacob) (passages quoted translated by C. MacKellar)

Reynier: et al. (2001), 'Mitochondrial DNA content affects the fertilizability of human oocytes', *Molecular Human Reproduction* 7: 425–9

Reynolds, J. and S. Fogel (2009), 'Letter to the New York stem cell research program ethics board,' *Centre for Genetics & Society*, 21 January 2009

Rideout III, W. M., et al. (2002), 'Correction of a genetic defect by nuclear transplantation and combined cell and gene therapy', *Cell* 109: 17–27

Robert, J. S. 2006 'The science and ethics of making part-human animals in stem cell biology', *The FASEB Journal* 20 (May 2006): 838–45

Robert, J. S. and F. Baylis (2003), 'Crossing Species Boundaries', *American Journal of Bioethics*, 3 (3): 1–13

Roetz, H. ed. (2006), *Cross-Cultural Issues in Bioethics: The Example of Human Cloning* (Series: *At the interface/Probing the boundaries* 27) (Amsterdam: Rodopi)

Rohleder, H. O. (1918), *Künstliche Zeugung und Anthropogenie, Monographien über Zeugung beim Menschen*, vol. 6. (Leipzig: Georg Thieme)

Rollin, B. E. (1986), 'On telos and genetic manipulation', *Between the Species* 2: 88–9

Rossant, J. and W. I. Frels (1980), 'Interspecific chimeras in mammals: successful production of live chimeras between *Mus musculus* and *Mus caroli*', *Science* 208: 419–21

Rossant, J. and A. Spence (1998), 'Chimeras and Mosaics in Mouse Mutant Analysis', *Trends in Genetics* 14: 358–63

Rossiianov, K. (2002), 'Beyond Species: Il'ya Ivanov and His Experiments on Cross-Breeding Humans with Anthropoid Apes', *Science in Context* 15 (2): 277–316

Ryder, R. (1971), 'Experiments on animals', in Godlovitch et al. 1971: 41–82

Rynning, E. (2009), 'Legal tools and strategies for the regulation of chimbrids', in Taupitz and Weschka 2009: 86–7

Sadr, R. al- (1986), *Al-falsafat al-ulya* (Beirut: Dar Al Kitab Al Lubnani)

Sample, I. (2005), 'Chromosome transplant in mice could provide clue to Down's syndrome illnesses', *Guardian*, 23 September 2005

—(2006), 'Stem cell experts seek licence to create human-rabbit embryo', *Guardian*, 5 October 2006

—(2009), 'Rival stem cell technique takes the heat out of hybrid embryo debate', *Guardian*, 13 January 2009

Sandoe: and N. Holtug (1993), 'Transgenic animals – which worries are ethically significant?', *Livestock Production Science* 36: 113–16

Saunders, R. and J. Savulescu (2008), 'Research ethics and lessons from Hwanggate: what can we learn from the Korean cloning fraud?', *Journal of Medical Ethics* 34(3): 214–21

Savulescu, J. (2003), 'Human-Animal Transgenesis and Chimeras Might Be an Expression of Our Humanity', *American Journal of Bioethics* 3 (3): 22–5

Schramm, R. D. and A. M. Paprocki (2004), 'In Vitro Development and Cell Allocation Following Aggregation of Split Embryos with Tetraploid or

Developmentally Asynchronous Blastomeres in Rhesus Monkeys', *Cloning and Stem Cells* 6 (3): 302–14

Schweizer, R. J. and H.-P. Bernhard (2009), 'National Law: Switzerland', Taupitz and Weschka 2009

Schwobel, C. and C. E. Gunton (eds) (1991), *Persons, Divine and Human* (Edinburgh: T. & T. Clark)

Science Watch (1989), 'Chromosome Transfer', *New York Times*, 25 July 1989

Scruton, R. (1996), *Animal Rights and Wrongs* (London: DEMOS)

Seller, M. (2008), 'Slipping on the slope of progress', *Tablet*, 5 April 2008

Shapira, A. (2009), 'Legal and Ethical Aspects of Chimera and Hybrid Research in Israel: Country Report with Comparative Observations', in Taupitz and Weschka 2009

Sheng, H.-Z. et al. (2003), 'Embryonic stem cells generated by nuclear transfer of human somatic nuclei into rabbit oocytes', *Cell Research* 13(4): 251–64

—(2008), 'Activation of human embryonic gene expression in cytoplasmic hybrid embryos constructed between bovine oocytes and human fibroblasts', *Cloning Stem Cells* 10(3): 297–305.

Shi, W. et al. 2003 'Epigenetic reprogramming in mammalian nuclear transfer', *Differentiation* 71: 91–113

Shinohara, T. et al. (2001), 'Mice containing a human chromosome 21 model behavioral impairment and cardiac anomalies of Down's syndrome', *Hum Mol Genet.* 10(11): 1163–75

Shreeve, J. (2005), 'The Other Stem-Cell Debate', *New York Times*, 10 April 2005

Singer, P. (1976), *Animal Liberation: A New Ethics for our Treatment of Animals* (London: Cape)

—(1979), 'Killing Humans and Killing Animals', *Inquiry* 22 (1–4): 145–56

—(1992), 'Embryo experimentation and the moral status of the embryo', in Matthews and Menlowe 1992: 81–91

—(2005), 'The Sanctity of Life', *Foreign Policy* (Sept/Oct 2005) 40

Smith, W. J. (2005), 'Is the world ready for a superboy?—or a dogboy?', *Dallas Morning News*, 13 November 2005

St John, J. and R. Lovell-Badge (2007), 'Human-animal cytoplasmic hybrid embryos, mitochondria, and an energetic debate', *Nature Cell Biology* 9: 988–92

Stephen, C. and A. Hall (2005), 'Stalin's half-man, half-ape super-warriors', *Scotsman*, 20 December 2005

Stoltzfus Jost, T. (2009), 'Legal Issues Involving Hybrids and Chimeras: United States Country Report', in Taupitz and Weschka 2009

Straughan, R. (1999), *Ethics, Morality and Animal Biotechnology*, (Swindon: Biotechnology and Biological Sciences Research Council)

Sulmasy, D. P. (2007), 'Human Dignity and Human Worth', in Malpas and Lickiss 2007: 9–18

Summers: M. et al. (1983), 'Synthesis of primary *Bos Taurus-Bos indicus* chimeric calves', *Animal Reproduction Science* 6: 91–102

Tachibana, M. et al. (2012), 'Generation of Chimeric Rhesus Monkeys', Cell 148: 1–11

Taupitz, J. (2008), 'The Chimbrids Project of the Universities of Heidelberg and Manheim', *Journal of International Biotechnology Law* 5 (3): 89–101

—(2011), 'Chimeras + Hybrids = Chimbrids: Legal Aspects', in Hug and Hermerén 2011

Taupitz, J. and M. Weschka (eds) (2009), *CHIMBRIDS: Chimeras and Hybrids in Comparative European and International Research* (Berlin: Springer)

Tecirlioglu, R. T., J. Guo and A. O. Trounson (2007), 'Inter-species Somatic Cell Nuclear transfer (iSCNT) and Preliminary Data for Horse-Cow/Mouse iSCNT', *Stem Cell Reviews* 2: 277–87

Tonti-Filippini, N. et al. (2006), 'Ethics and Human-animal Transgenesis', *National Catholic Bioethics Quarterly* 6(4): 689–704

Tooley, M. (1983), *Abortion and infanticide* (Oxford, UK: Clarendon Press)

Václav, O. et al. (2001), 'Segregation of Human Neural Stem Cells in the Developing Primate Forebrain', *Science* (7 September 2001), 293(5536): 1820–4

Van Steendam, G. et al. (2006), 'The Case of Reproductive Cloning, Germline Gene Therapy and Human Dignity', *Science and Engineering Ethics* 12

Vercors (1952), *Les Animaux denatures* (trans.) Rita Barisse (1954), as *Borderline* (London: Macmillan)

—(2009), *Zoo ou l'Assassin philanthrope* (Paris: Magnard)

Verhoog, H. (1992), 'The concept of intrinsic value and transgenic animals', *Journal of Agricultural and Environmental Ethics* 5: 147–60

Vogel, G. (2006), 'Stem cells. Team claims success with cow-mouse nuclear transfer', *Science* 313: 155–6

Voltaire (1734), *Lettres philosophiques* (passages quoted translated by C. MacKellar)

Vorstenbosch, J. (1993), 'The concept of integrity: its significance for the ethical discussion on biotechnology and animals', *Livestock Production Science* 36: 109–12

Wallace, D. C. (1999), 'Mitochondrial diseases in man and mouse', *Science* 283: 1482–8

Wang, Z. and R. Jaenisch (2004), 'At most three ES cells contribute to the somatic lineages of chimeric mice and of mice produced by ES-tetraploid complementation', *Developmental Biology* 275 (1): 192–201

Warren, M. (1994), 'On the moral and legal status of abortion' in Beauchamp and Walters 1994: 302–11

Waters, B. and R. Cole-Turner (eds) (2000), *God and the Embryo: Religious Voices on Stem Cells and Cloning* (Washington D.C.: Georgetown University Press)

Watt, H. (2009), *Incapacity and Care: Controversies in Healthcare and Research* (Oxford: The Linacre Centre)

Watts, G. ed. (2009), *Hype, hope and hybrids: Science, policy and media perspectives of the Human Fertilisation and Embryology Bill* (London: Academy of Medical Sciences/ Medical Research Council / Science Media Centre/ Wellcome Trust)

Weiss, R. (2004), 'Of mice, men and in-between: Scientists debate blending of human, animal forms', *Washington Post*, 20 November 2004

—(2005), 'U.S. Denies Patent for a Too-Human Hybrid', *Washington Post*, 13 February 2005

Wells, H. G. (2005) (1896), *The Island of Doctor Moreau*, (London: Penguin Classics)

Wen, D. C. et al. (2005), 'Hybrid embryos produced by transferring panda or cat somatic nuclei into rabbit MII oocytes can develop to blastocyst in vitro', *Journal of Experimental Zoology Part A: Comparative Experimental Biology*, 303A (8): 689–97

Wheale: and R. McNally (eds) (1990), *The Bio-Revolution* (London: Pluto)

White, D. G. (1991), *Myths of the Dog-man* (Chicago, IL: University of Chicago Press)

Williams, B. and A. Sen (eds) (1982), *Utiliarianism and Beyond* (Cambridge: Cambridge University Press)

Williams, G. (1958), *The Sanctity of Life and the Criminal Law* (London: Faber and Faber)

Williams, T. J. et al. (1990), 'Production of interspecies chimeric calves by aggregation of *Bos indicus* and *Bos Taurus* demi-embryos', *Reproduction Fertility and Development* 2: 385–94

Wilmut, I. et al. (1997), 'Viable Offspring Derived from Fetal and Adult Mammalian Cells', *Nature* 385: 810–13

Winnett, R. (2008), 'Embryo bill: David Cameron voted for human-animal hybrids over disabled son', *Telegraph*, 20 May 2008

Wittgenstein, L. (1969), *On Certainty*, edited by G. E. M. Anscombe and G. H. von Wright, trans. Denis Paul and G. E. M. Anscombe (New York: J. and J. Harper)

Wood, C. and A. Westmore (1984), *Test-Tube Conception* (Sydney: George Allen & Unwin)

Wyatt J. (2009), *Matters of Life and Death* (Nottingham: IVP/CMF)

Yokoo, T. et al. (2005), 'Human mesenchymal stem cells in rodent whole-embryo culture are reprogrammed to contribute to kidney tissues', *Proceedings of the National Academy of Science of the United States of America* 102 (9): 3296–3300

Yang, C. X. et al. (2003), 'In vitro development and mitochondrial fate of macaca-rabbit cloned embryos', *Molecular Reproduction and Development* 65: 396–401

—(2004), 'Quantitative analysis of mitochondrial DNAs in macaque embryos reprogrammed by rabbit oocytes', *Reproduction* 127: 201–5

Zizioulas, J. (1991), 'On Being a Person: Towards an Ontology of personhood' in Schwobel and Gunton 1991

Index